NORTH IDAHO
MW01624031
Estep Cabin
Ramey
BM 83
MINE 8551
McFadden Point
3212
Prospect
Copper Camp
Creek
Ford
Monumental Bridge
Monumental Bar
Carpenters Gulch
Diamond Point
COPPER 7981
Diamond Creek
BM 4751
BIG 8500
FRANK CHUR
WILDER
Slide Gulch
BM 4914
9323
Marble Mountain
Center Mountain
9104
Snowslide Peak
MOUNTAINS
Simonds Landing Strip
BM 8713
BM 7544
Basin
Snowslide
Mud
Bear Lake
Catherine Lake

Two-Man Stick

Memoirs of a Smokejumper

Two-Man Stick

Memoirs of a Smokejumper

Bud Filler

Marianne,

Your brother asked that I send you this book. I hope you enjoy the jump stories and the good 'ol days in the Idaho mountains.

Very best wishes

Bud Filler

Mar 7/2008

Burning Mountain Press
Boise, Idaho

Published by Burning Mountain Press
P.O. Box 45534
Boise, Idaho 83711-5534

Library of Congress Cataloging-in-Publication Data

Filler, Bud, 1934-
Two-man stick: memoirs of a smokejumper / Bud Filler.-
1st ed.
p. cm.
ISBN 1-891055-02-1.-ISBN 1-891055-01 (pbk.)

1. Smokejumpers. 2. Idaho. 3. U.S. Forest Service

1999
Printed on acid-free paper by Western Printing, Lewiston, Idaho
Designed by Dennis Held
First Edition 1999

10 9 8 7 6 5 4 3 2 1

To the men and women of the United States Forest Service Smokejumpers, the tough and the elite.

Acknowledgments

First, thanks to Ellie, my wife, who has listened to my jump stories and tales of the forest for years. After attending the smokejumper reunions in McCall and Missoula, she now believes that some of these stories may actually be true. She said they should be recorded for the children and the grandchildren. Others have said the same; few people know what we did in those days of smokejumping in the mountains of the West. It was a fleeting time in many of our lives, involving only several summers for some, but the friendships and adventures remain in our memories forever.

Names of the lookout operators in several of the chapters are fictitious. The names of all the other characters are real, with the exception of Liz Holden, whom I have not seen for decades, and Len Walker and Frank Kerby. Their names were changed to protect them from embarrassment, and, possibly, glory.

I would like to express my appreciation to the following for their help in reading the chapters and providing inspiration to keep typing: Del Catlin, Earl Dodd, Dub Horn, Reid Jackson, Ted Koskella, Jim Larkin, Georgina and Tim Larson, Verna Peterson, Carl Rosselli, Natasha Roth, Gordon Saxon, Linda Stedman,Claudia Vant Hof and Wayne Webb. Thanks to Neal Davis and Scott Anderson of the USFS Smokejumper unit at McCall, Idaho, for providing access to the archives. Dr. Doug Cords, from Fresno State, read the original story and made valuable suggestions. Thanks, Doug.

Many of the details of the aircraft were taken from Steve Smith's book *Fly the Biggest Piece Back,* and from William T. Larkin's writings in *The Ford Tri-Motor.* The dates of the first smokejumper activities and quotations are from Stan Cohen's *A Pictorial History of Smokejumping.*

A special tribute to my mother, who saved the letters and read the drafts; and to my father, who has always given me good advice.

Dennis Held edited the manuscript, and I thank him for his help.

And last, I want to express my appreciation to Wayne King, hot-shot firefighter from the Angeles National Forest, logger, business partner, long-time friend and fellow pilot. He said nothing was hotter than a California fire, when the devil winds came in from the east and the chaparral exploded like gasoline. I told him he should have tried to stop a running crown fire in the lodgepoles. It was stories like these around a thousand campfires that got me started.

PHOTO CREDITS: USDA Forest Service, Payette National Forest: all color photographs except as noted; front and back cover background photographs, all black and white photographs except page 21, which is courtesy of Jim Larkin; cover inset and page 130 are courtesy of Ray Beaseley.

Contents

The torture rack: smokejumper training in the fifties

STICK: *Any number of parachutists dropping together.*

Webster's Dictionary

Two-Man Stick

Memoirs of a Smokejumper

First fire jump out of McCall, 1943

Prologue

Fighting forest fires by dropping men from the skies by parachute was first started on an experimental basis by the United States Forest Service in the late thirties. The idea was to get to the fires when they were small, put them out and save from destruction huge tracts of timber in the inaccessible regions of the West. To carry out this new endeavor, several skilled parachutists and woodsmen started training at the Forest Service station at Winthrop, Washington. One of the first jumpers, J. B. Bruce, demonstrated that parachuting techniques could, indeed, be used to take firefighters to blazes in remote wilderness regions.

There was opposition from the top echelons of the Forest Service to this concept. In 1935, regional forester Evan Kelly expressed displeasure with the idea of aerial firefighting in a letter to Earl Loveridge of the Forest Service in Washington, D. C.: "I am willing to take a chance on most any kind of proposition that promises better action on fires, but I hesitate very much to go into the kind of thing that Bruce proposes. In the first place the best information I can get from experienced fliers is that all parachute jumpers are more or less crazy, just a bit unbalanced, otherwise they wouldn't be engaged in such a hazardous undertaking; accordingly, I discount materially the practicality of Bruce's ideas."

But smokejumping advocates prevailed, and the training and testing continued. In 1940, Earle Cooley and Rufus Robinson made the first successful parachute jump on a lightning-caused blaze in the mountains of the Nez Perce National Forest in northern Idaho. These men, along with others who followed, proved that those who accepted and even enjoyed the dangers of parachuting could suppress wildfires before they consumed thousands of acres of timber.

Today, some voices cry out to proclaim a forest fire an act of nature, part of the chain of events in the forest. Let the timber burn, they say, so it doesn't get thick and catch fire later. They say this even during a fire, when the trees disappear as ash in the sky and the forest floor is scorched to the mineral soil. Trees that grew for 250 years become charcoal -pointed spears in the draws and on the ridges. Charred hillsides, once covered with timber, foliage and flowers, give way to grass and weeds and bare slashes of erosion. For centuries, fires did occur in the forest, but they were limited, and that was before man came upon the scene. The natural chain of events would more likely follow the idea that the trees just might be allowed to mature, grow tall and old, and fall onto the forest floor to provide an environment for the next generations.

It was our challenge as firefighters to control the natural disaster of fire in the remote wilderness country. The large destructive fires, which consumed thousands of acres of timber and required thousands of men, were to be put out while they were small. That was the goal of the Forest Service, and that is what we did.

The events in this book took place in the summers of 1952, 1953 and 1954, with the story about the coyotes and the firefighters from the fire season of 1956.

Very little has been recorded about those early years. We were all young then and so was the concept of fire control from the sky. In a classroom training session at the McCall, Idaho, unit we watched the 1950 film, *Telephone Creek*. The ranger in the movie helplessly watched the flames of a forest fire. He said, "I don't like to see the baby trees burn." Coughs and snickers were heard in the room. But, thinking back, I believe we were as dedicated as that ranger.

This story is about the endless ridges of green timber and granite peaks under the silver wings of a plane, and of the wild and windy places of the earth where few people had traveled. It is also the story of a young fire, when smoke begins to curl upward between the tall spruces, and of the explosive yellow-red of a crown fire as it roars to the top of a ridge. This book is about a group of young men, tough and dedicated, who lived, loved, and continued to return to the mountains. These men, with their heavy canvas suits and parachutes, climbed aboard the Ford Tri-Motors and Travelaires, and flew over the mountains into immortality. They called themselves smokejumpers. I consider myself fortunate to have been one of them.

Carl Rosselli spotting on a fire jump

Chapter One

WILD COUNTRY

You who love the haunts of nature,
Love the sunshine on the meadow,
Love the shadow of the forest,
Love the thunder in the mountains,
Whose innumerable echoes,
Leap and bound from crag to cliff-side,
Listen to these admonitions,
For the guidance on your journey.

The Song of Hiawatha
Henry Wadsworth Longfellow

If I had been looking I would have seen them, but my mind was on other things, like surviving the jump and not getting broken up in the rocks.

They were seven sleek mountain goats and they came down the rocky chute from the top. The goats stopped on a ledge over a drop of broken boulders. A vast wilderness fell off below them.

This was their home, the high wind-swept ridges above the canyon of the Middle Fork of the Salmon River.

Alpine firs and alders grew at timberline at the heads of the draws below the granite slides. Small trees gave way to the Douglas firs, and down on the lower slopes the stately ponderosa pines stood, their orange bark visible on the sunlit days from the peak. Miles below, the river coursed its way northward.

An alpine lake was hidden in the rocks below the peak. The secret place was a favorite haunt of the goats, where they visited to feed on the grasses along the edges. At the outlet end of the blue water, tall spruces guarded a brook as it plunged into the valley. The breezes coming up from the river canyon were hot and the dark branches of the spruces swayed in the afternoon sun. Lower on the mountainside the needles of the Douglas firs turned silver as they rolled to the changes in the wind.

The big white billy, largest of the goats, weighed over 300 pounds and had seen many summers. His hooves and horns were as dark as polished ebony. The big goat had been watching the dark clouds to the west, the distant flashes of lightning glistening in his eyes. The wind rippled the clots of wool hanging below his belly and blew dust past his hooves.

The rains came, the drops drying quickly on the rough surfaces of the rocks. Lightning flashed in the peaks and thunder cracked along the ridges. Blasts of hot air coming up from the dry canyon mixed with the cold air in the mountains, and the trees on the ridges seemed to swirl, disappearing and reappearing in the clouds.

The big goat turned and moved his family to a sheltered crevice in the rocks. The wind and the lightning would be gone in the morning and he would lead his band down to a meadow in a mountain pass where a spring bubbled in the patches of mountain maple. Fresh grass and shrubs grew around the spring. In his life the big billy had seen many storms crashing onto his mountains. But he had never seen men. Before the sun set again, he would see them and they would come from the sky.

A mile away, a large Douglas fir grew in a forest opening, just off the crest of the mountain. The tree was an old giant, tall and heavily branched, its lower limbs thicker than the trunks of most trees on the mountain. Beneath the tree, the ground dropped away into a tangle of mountain cherry. Near the fir, stands of lodgepole grew, straight and tall, and the pines blended into a forest of old-growth firs.

The needles of the Douglas fir were soft and swayed easy in the mountain breezes. The summer had been dry and vapors from the resins in the needles floated away in the warm winds.

The bark of the tree was four inches thick at the base of the trunk. Over the years, lightning had scarred the Douglas fir from top to bottom. The tree had survived two centuries of storms and fires. It would not survive the night.

The lightning came with the darkness. A flash from the clouds ripped into the branches, met a charge from the ground, and in a brilliant white microsecond the trunk of the giant exploded. A section of wood burst from the trunk and somersaulted in the air, returned to earth and buried itself in the soft dirt. The shattered top of the tree flew with the wind to the ground. From the broken top, flames erupted and a white cloud of smoke rolled up and over the ridge.

Below the limbs and protected from the rains, a carpet of bone-dry twigs and needles had accumulated. This forest litter ignited within seconds. With the rush of the night air the leeward side of the tree became a stepladder of yellow flames from the ground to the wind-tossed top.

The cones of the Douglas fir were two inches long and almost as round. Nature designed them to roll downhill and come to rest against a log or fallen branch. The cones would also roll when on fire. Flaming and smoking, they bounced their way downhill.

The storm moved into the far peaks. The fire that flashed and burned the side of the giant tree retreated to the ground. The broken top continued to smolder and white smoke rolled with the wind. As the night grew darker a line of flames, like a dozen campfires, crept downhill into the dry tangle of wild cherry.

❖ ❖ ❖

Fifteen miles to the northwest a first-year forestry student stood on a wooden box in the center of the plank floor of a fire lookout. Built of timbers and standing seventy-five feet tall, the lookout tower had been built on one of the highest peaks in the mountains of the Middle Fork country. It was dark now, the wind rattled the glass in the windows, and a blue light, generated from the lightning-charged atmosphere, circled the room like a pale ghost.

Pete, the forestry student, had counted seventeen lightning strikes and was recording the compass bearing to each. The ink-black night, the lightning and the thunder crashing against the tower became distracting and Pete began to miss some bearings. Another flash on a distant ridge, and he swung the alidade to register the bearing. The alidade, a rotating device mounted on a table in the middle of the tower, could be turned to determine the direction to any distant point. Pete was scrambling to work and still remain upright on the wooden stool.

Out there! Flames! Fire from that hit! Pete concentrated on where he had seen the flash of fire. *It's dying down. I can't see a thing. It's dark again. This might be a sleeper.* A sleeper was a fire that consumed some branches and trees, then went to ground, slowly burning the pine needles and flaming up later when it reached more fuel and got more oxygen, often after a week or more had passed. *I've got the bearing. I'll watch for flames during the night. Better report this one to George at the guard station. The storm probably blew trees down across the telephone line. I'll use the radio.*

"Bear Point to Grouse Creek," Pete called, urgency in his voice.

"Go ahead, Pete." George, the assistant ranger, was waiting, expecting a call from the Bear Point lookout operator.

"George, I've recorded eighteen strikes. The last one flamed up real good, but the fire has died down."

"Give me the bearing and the estimated distance and location, Pete," George said. "And how are you doing up there?" Without waiting for a reply, George continued, "That lookout tower has taken bigger storms than this one. Just stay on the wooden stool. I'll call Horse Heaven lookout to see if Dade has a bearing on that strike. That storm went right across your country. Check that location again during the night."

Pete's bearing was probably right, but exact distances were hard to judge and parallel ridges could blend together and appear to the eye as one large ridge. Thick forests of lodgepole and fir grew in the high country above Pistol Creek: it was no place to get a fire going.

George turned the crank on the telephone. He used the field phone to determine if it was working and to keep the radio open for calls from other lookouts. "Hello, Dade, this is Grouse Creek."

"George, I've got three smokes," Dade said. "One is down in Pete's direction. Looks like the main ridge over Pistol Creek. The bearing is 224 degrees." It was Dade's third summer on the lookout tower and he was already an old hand.

"Yeah, Pete also reported that strike and some flames. Keep track of the other two smokes. Give me their bearings in ten minutes." As he was talking, George extended the compass directions from the two lookouts on the table map in front of him. He noted where the two lines crossed to find the exact location. The Pistol Creek country near the breaks of the Middle Fork was two days away by pack string and ground crew.

"Call me at first light," George said. "I'll call Payette fire dispatch. You can expect some aerial company at Pistol Creek at dawn."

"Thanks. I'm still busy here--this storm isn't over yet. Horse Heaven out." Dade picked up his binoculars and scanned the dark mountains.

The mountain goats, their short wooly tails to the wind and their black eyes now merely slits, lay in the lee of the rocks of the peak, waiting for the morning light.

Leo Brown, Bruce Yergensen, Bob Thornton and Sid Root in the Ford

Chapter Two

THE MEN

"It is not enough to say we will do our best; we must do what is necessary."

Winston Churchill

It all began in the spring of 1952, when I told my mother and father I was going west to work on a fire lookout tower for the United States Forest Service, somewhere in the mountains of Idaho. I felt bad stretching the truth to them, but I had received a letter from Jack Kooch, supervisor of the Payette National Forest at McCall, Idaho, offering me a job as a smokejumper. Most of my friends didn't know what that meant, but I did. I'd been reading about this new group of firefighters who parachuted from the sky to put out forest fires in the remote mountains of the West.

The year before graduation I had asked my dad for permission to join the Marines, but without even looking up at me over the newspaper, he said no, adding that he would break both my arms if I asked him again. That matter was closed. I was seventeen and he probably would have done it. As usual he was right; several of my high-school friends volunteered, went to Korea, and I never saw them again.

I was enrolled in forestry at Penn State in the fall and I wanted to get a job in the woods, any kind of outdoor work. Getting hired on as a smokejumper was beyond my wildest dreams. I had written to Dr. Maurice Goddard, the dean of the forestry school, who recommended writing directly to Mr. Kooch. Maybe it was a little of "who you know" even then. A new unit of jumpers had been started at McCall in 1947, and Mr. Kooch wanted to build on this group. My resume was about average, I thought, and it didn't show that I was a mountain man, French trapper of the north woods, ex-paratrooper from the Foreign Legion or anyone like that, just a guy in forestry who wanted a challenging job. The return letter from Mr. Kooch was in my left shirt pocket and I guarded the envelope with my life.

But there was another problem. Somewhere during the last year I had acquired some serious, painful calluses on my feet. I sharpened a hunting knife to a razor edge, sterilized it over a candle flame, and proceeded to cut away. The calluses bled like mad and immediately got worse. It was a miracle that blood poisoning didn't set in. So off to a doctor I went, who said surgery was required, said I'd be on crutches for six weeks, and I said no thanks, knowing that could ruin the whole summer.

My mother said go to a dermatologist. The dermatologist saved me. He said those were in-grown warts, and warned me not to carve on them with a bowie knife anymore, and to dust my feet with a strong-smelling yellow powder, take the pills he prescribed, and the problem would go away in six weeks. But in the meantime, walking, running and jumping would be painful, which I already knew. Later I found that landing flat-footed from a jump tower or from the sky was like having red-hot needles shoot all the way from my feet to my stomach. But I didn't tell a soul, particularly the squad leaders, for fear of not qualifying for the group. Once in a while somebody would ask me why I was limping, and I'd say it was nothing. The doctor was right; in six weeks the warts were gone.

If you were short on funds and didn't have a car, which was my situation, the best way to get from Pennsylvania to Idaho was by train. Again, my dad came through and got me a pass for $25, since he worked for the Pennsylvania Railroad; reciprocal travel for family members on other railroads was a benefit, like people get by working for the airlines today. For $25, I got a coach seat all the way, three days of sitting up and listening to the clack of the steel wheels over the rails. Another $10 for cheese sandwiches and I knew I could make it to Idaho.

Jim McLinn, another jumper from Boise, told me later he had only one pack of cigarettes in his pocket when he was discharged from the Air Force and he made the trip from Chicago to Boise on those smokes.

Four days later I stepped from the passenger car onto the brick platform at the Union Pacific depot in Boise, Idaho. The air was dry and fresh and the sky was blue and clear. The city of Boise was beautiful, with tree-lined streets reaching up to the hills to the north. But where were the forests? The rolling hills were bare.

Inside the depot, the stationmaster directed me to the bus terminal, a walk of a mile to the center of town. I threw my duffel over my shoulder and started down Capitol Boulevard. Stop one block short of the Capitol Building and go left for three, he said. The train ride had left me groggy and the exercise felt good. I overlooked the pain in my feet.

Panting, carrying the heavy bag down Capitol Boulevard, I looked to the north and a thin line of trees came into focus above the dry hills. Yes, the forests were up there somewhere. I learned they were there all right and they stretched from the mountains north of Boise to the Canadian border, and beyond.

The ticket man behind the window at the Boise-Winnemucca Stage bus terminal took most of the last of my money, leaving just enough to splurge on a hamburger and a cup of coffee. He gave me the change in silver dollars. Silver dollars, heavy in my jeans pocket: I was in the West at last. When I got to McCall I was determined to make the grade, and in a week or two a paycheck would be coming from the Forest Service. It *had* to work that way, because that was the last of my money. The station attendant said the bus ride from Boise to McCall took two hours. (I wondered for years, where in the hell was Winnemucca?)

I climbed aboard and sat behind the bus driver so I could ask him lots of questions. He seemed glad for the company and answered every one, telling me about the mountains, the whitewater rivers, the high passes. The only other passengers were three elderly ladies. Visiting their relatives in Grangeville, they said. They also said I would see lots of timber. Thirty miles north of Boise the first trees appeared, big ponderosa pines along the winding Payette River. The river was named for a French trapper, the driver added, who explored and trapped in this country long before Lewis and Clark came through in 1805. For another thirty miles the road curved and climbed past whitewater rapids and steep canyon walls. Just when I thought this beautiful wild river would go on forever, the road came out of the timber into a wide open valley, lush with spring grass. The blacktop went straight north, through pastures and pines, until we arrived at a beautiful mountain lake and the village of McCall.

I stepped off the bus at the local café, shouldered the duffel, and walked west along Lake Street in the direction the bus driver said would take me to the forest supervisor's office. The water of the lake on my right was deep, a blue-purple, and a light wind rippled the surface. A few small stores and saloons were squeezed in between Lake Street and the water. The May afternoon was warm, the air dry and clear. The high cloudless sky was everywhere. I could smell the pines. Timbered mountains stretched on forever. The Appalachian mountains where I roamed in Pennsylvania were high and rolling, but this was something else entirely, a Shangri-la. At that moment I was determined that I was going to be a part of this outfit.

After I walked a mile I saw an official sign at the edge of a well-tended lawn: PAYETTE NATIONAL FOREST—UNITED STATES FOREST SERVICE. The building in the lawn's center was painted a dazzling white with green trim. Behind the office was a large dusty parking lot with barn-like warehouses on its east and west ends. Four government pickups were parked just to the back. There's a comfortable feeling here, I thought, walking through a tunnel of spruce trees up to a doorway and a sign that read FOREST SUPERVISOR.

Maybe I'd make a career of the Forest Service. Might even be a ranger some day. Now that would be a challenge. I *was* enrolled in forestry and that had helped me get the job. I tapped the letter in my shirt pocket when I opened the door.

An attractive woman at the receptionist desk in the supervisor's office said that the fire season had not yet begun, and to report to the smokejumper loft just over the hill. "Follow the dirt road, you can't miss it," she said. She looked at me carefully, gave me a smile that lasted two seconds longer than I thought it should for a woman in her high position. I smiled back and said thanks.

On the road again, I shouldered the duffel. The afternoon quiet was broken by the squeal of rubber on blacktop and a clank of metal as a beat-up Jeep bounced around the corner from Lake Street and headed up the dirt road toward me. The canvas top had been removed and the windshield was folded down and snapped to the hood. The Jeep rumbled up and stopped. Two young, suntanned men in the front seat were obviously having a lot of fun. They both had wide grins on their faces as they looked me over. Their outfit was heavy logging boots, faded denims and white T-shirts. The driver had a pack of cigarettes rolled up under his left sleeve, and his passenger had his feet braced against the battered dashboard. These two guys were trim and looked pretty tough. I felt a little wan after spending the last four days on a train.

"Going to the smokejumper loft?" the driver asked.

"You bet. This the way?"

"It's the way. Going to be a jumper?" They were still grinning at me, like maybe they knew something I didn't.

"I'm Gene Marker," he said, "and this is Carl Shaver." I introduced myself and we shook hands. Marker had short, light-brown hair and mischievous blue eyes. Carl Shaver had coal-black hair and large, dark eyes. They had the look of having been around McCall for years, maybe even seasoned woodsmen.

"Hop in the back. We'll drop you off at the loft. Sign in at the desk," Marker said. They said they were first-year men too, and they had gotten out of school early and were working out every day to get in shape for the regimen to come. I quickly learned that the McCall smokejumper unit of Region 4 of the Forest Service was a small, tough, and very selective group of men. I was amazed that they had pulled my application and picked me, and I glanced upward and said thanks to the Big Man upstairs. From what we were going to be doing, I thought it might be wise to be on His good side.

After checking in at the office, I looked around the facilities. Behind the loft were an athletic field, a jump tower, an obstacle course and various pieces of equipment, some resembling medieval torture racks. And that is what they were actually called. Two gray-painted wood bunkhouses and the cookhouse squatted across the dirt road from the loft. Somebody assigned me a bunk and a locker not far from the wood stove. It was a good location: in late May, the mornings were cold in the mountains and snow was still falling in the high country.

That evening the old jumpers, the squad leaders and the new guys all began gathering in the bunkhouses for introductions. I considered the old jumpers anyone on their second season or more. The squad leaders had many summers under their belts and many of them had year-round appointments with the Forest Service. The old pros looked over the new men, and the new men wondered if we could make it. Then the stories started—about the planes, the jumps and the fires—and the tales of barroom adventures. Any new man with a faint heart could have dropped out at that point. No one did.

I learned that the primary aerial firefighting base in the Rocky Mountains was at Missoula, Montana. From 1943 through 1947, the Idaho jumpers trained at Missoula and then transferred to McCall to cover fire control in Region 4. In 1943, five jumpers started in McCall, making a total of fifteen jumps on five fires. The jump planes and pilots were operated by the Johnson Flying Service out of Missoula, and dispatched to Idaho. After 1947, the McCall unit became an independent training and jump base. By 1952 the unit size was increased to forty-seven men, with another twelve operating, after training, out of Idaho City. That year twenty-eight "Neds"—the name coined later for first-year men—began training. Of the new guys, six of us were eighteen years old. Looking back on the statistics, I'm not sure how I got selected, but I was glad I was.

On the first day of June, we were assigned to squads and sent to the ranger stations for project work. This was the start of physical conditioning for the jump training and the fire season to come. We reopened trails in the forest, fixed up lookout towers, repaired telephone lines and rebuilt pack bridges across rivers. In addition to the forestry maintenance work, we did calisthenics in the mornings and evenings. The squad leader's cadence was sharp in the cold air: "Up one, up two, up three, gimme another twenty-five, and I want them now." Heavy coats and gloves were standard until noon, when the early summer sun warmed the woods. Cobwebs from the college books and off-season jobs broke away in the first week and in the second week sore muscles began to heal. We did our own cooking in the ranger stations and guard cabins. We ate well and our muscles began to build. The cold air, project work, physical exercise and healthy meals burned away the softness from the winter's debaucheries.

After two weeks we returned to the McCall unit and the training got serious. Besides the "daily dozen" exercises, there were workouts on the torture rack: backbends with the arms folded behind the head and the stomach muscles stretched beyond the limit. We hit the obstacle course, climbed ropes, and ran two miles wearing the six-pound logging boots. Army basic training, I found out later, was a sneeze compared to this. The selection process had been thorough and, of the thirty-plus young men chosen, only a few didn't measure up. Once into training, some realized that stepping out of airplanes simply wasn't a rational thing to do. Some had attitudes that couldn't adapt to the program, or to the others. Some new men—as well as a couple of the veterans—broke ankles and legs on the practice jumps, and their careers, unless they were very persistent, were over.

The age range was from eighteen to twenty-eight, although the maximum age was later raised. The weight limit was 130 to 180 pounds. Whether this was with the six-pound jump boots on or buck naked was never clearly defined, but eyes were turned aside for a few returning big guys who had picked up some extra pounds during the winter. The height restriction was between 5'3" and 6'3", but few jumpers were under 5'6". Other requirements were some experience in firefighting and skills in woodsmanship, and we had to pass standard Civil Service physicals.

The second-year and older jumpers made three practice jumps and called themselves ready for the fire season. The new men made seven practice jumps, first over flat open meadows, and then on smaller and smaller landing zones.

For the new men, the four-week training period was almost over. The next-to-last phase was leaping off a fifty-foot wooden tower. We suited up and climbed to the top. A rope was attached to the D-rings of our canvas harness. The other end was tied to a set of springs and a cable, which stopped our descent a few feet from the ground. This was meant to simulate the opening shock of a parachute in the air. It was a weak simulation at that, I later learned, because when we actually jumped from a plane, our bodies were in all kinds of different positions, because of the force of the wind whipping by, so the opening snap of the chute in midair was actually more like getting cracked at the end of a long whip. From the tower, looking out from the framed door at the top to the stretched canvas safety net below gave my stomach a few queasy loops, more scary than looking down from the open door of a plane in flight, 2,000 feet in the air. Hard to figure.

The next morning we climbed aboard the Ford Tri-Motor for our first jump. We would go out one at a time, a squad leader checking our equipment and our mental attitude as we shuffled, crouched over from the canvas harness, to the open cabin door. Another squad leader was on the ground with a bullhorn, shouting instructions to us as we fell to the meadow—as if we had any other place to go. Show no fear, I thought. I remembered when, as a kid in Pennsylvania, I'd jumped off the old wooden Huntington and Broadtop Railroad trestle into the Juniata River. I had to—my friends were watching.

This was my first time in a plane, and I would be getting back to the ground from the air. I'd do this ten or eleven more times before actually being seated in an airplane when it landed.

We lifted off and climbed to the jump altitude of 1,000 feet. North of town, Payette Lake shimmered bright and blue in the clear morning air, with several white streaks on the water from early morning speedboaters. Rich guys with boats and waterskis, I thought. I'd rather be up in a noisy plane, with a parachute strapped to my back.

The meadows to the south sparkled with a touch of dew. Patches of fog were suspended like small white balloons above the pine trees along the river. The day was going to be a warm one. Two Forest Service pickups were parked at the edge of the meadow and somebody put out a tiny orange target in the middle of a green field.

The squad leader threw out the drift chute. We were trained to observe any wind drift, starting from the first jump. This morning the winds were calm, just the way the leaders planned.

I was third in line. The first two guys departed the door after a healthy slap on the back from the spotter. Now it was my turn. The Ford banked for the jump pass, the low wing allowing the morning sun to splash through the cabin windows across the faces of the other five jumpers, who sat on the floor, awaiting the nod from the spotter. They looked tanned, ready for anything, but somber. Maybe I did too. I was holding onto to the sides of the door, trying to keep upright. No seat belts were installed in the cabin because there were no seats. We wouldn't have used them anyway.

The pilot leveled the wings. This was it. I made an equipment check for the third or fourth time, then snapped my static line to a cable just inside the door. The spotter made another safety tug on my harness. "Keep your eyes on the horizon when you step out. Don't look down," he shouted. I took a big gasp of air—just like on the old railroad trestle—sucked in my stomach, felt the slap on my back, and stepped into space.

The only thing I remembered later about the first jump was the swirling of the high blue sky and the clouds. Then came the opening shock, more abrupt than I expected. When the swinging stopped, I heard the bullhorn: "Pull on the right guideline. Pull on your left guideline. Pull down on both risers. Keep your feet together." A small cluster of people were down there, looking up. There was the orange marker. I wasn't going to be close. Maybe next time.

"Keep your feet together!" the bullhorn roared. "Together!"

Bang. Where did that ground come from? I tried to do an Allen roll, but it wasn't very pretty. As the dust settled, I stood up and was suddenly very happy. Nothing was broken. Ray Beasley and Bob Donnelley came running over, slapped me on the back, offered congratulations, then helped me roll up the white parachute.

After that we made two jumps a day. It is said the second and third jumps were the worst, because you knew what to expect. But after that things got easier, even fun, at least when the conditions were perfect. But they were almost never perfect.

On the sixth and seventh jumps we made tree landings, to practice the let-down procedures we learned, or should have learned, if we came down in a tree on an actual fire jump, which we frequently did. We carried 100 feet of half-inch cotton rope in a lower leg pocket of our jump suits. When suspended in a tree, we threaded the rope through the D-rings of our harness and tied off to the parachute risers. The jumper made a loop to stand on, to release the weight of the body, unhooked the D-rings of the chute, and rappelled down the rope. It worked well in practice and most of the time on actual tree landings. On a fire jump once, Carl Shaver wound up twisted head-down in the branches of a tree and had to be inventive, but he made it.

On actual fire jumps, tree landings were made to avoid rocks or leg-breaking timber blow-downs, but sometimecrosswinds or poor parachute control took the jumper into the trees by accident. Crashing through the branches and coming to a springy stop near the top of a 150-foot fir on the downhill side of a sloped canyon made you want to remember all you'd learned in training about let-down procedures. And the rope was only 100 feet long.

The second four-week phase of training developed our skills in nine categories: history of the Forest Service and ethics on and off fires (don't get into too many fights); parachuting procedures and planes; smokechasing; cargo retrieving; fire suppression; fire escape; first aid; communications; and more physical conditioning.

Most of the fires put out by aerial control were small, started by lightning in the back-country and manned by jumpers within hours, or even minutes. The intent was to reduce suppression costs and prevent fires from becoming too large and dangerous. Once the fires blew up and went on their own across the mountains, they destroyed thousands of acres of timber, and sometimes claimed the lives of smokejumpers, hot-shot firefighters, ground personnel and aviators.

After six weeks of training, fifty-nine men remained at

McCall. Twelve were transferred to Idaho City, Idaho, where a base was located near a dirt airfield at a ranger station of the Boise National Forest. Together the Idaho firefighters would cover eight National Forests. When needed, the Idaho jumpers flew into Montana, Wyoming, Oregon, and New Mexico, and in later years, north to Alaska.

In 1952, the squad leaders, all experienced jumpers, were Lloyd Johnson, foreman; Reid Jackson; Del Catlin; Wayne Webb; Richard A. Peterson; Seymour Peterson; Smokey Stover; Robert Caldwell; Francis Doffing; Jack Wilcox; and Carl Rosselli.

Smokey Stover, Francis Doffing and Jack Wilcox operated from the Idaho City base. Lloyd Johnson was the McCall foreman in 1952 and was replaced by Reid Jackson in July of 1953. The men didn't want to see Lloyd leave, and even signed a petition to extend his service. It was to no avail. Reid became the new boss and gained as much respect from the men as they gave to Lloyd, and Reid became leader of the pack.

During the training jumps, four new men broke ankles or legs, leaving twenty-eight to start the fire season. Twenty stayed at McCall, and the other eight volunteered, or by luck of the draw, moved to Idaho City. The new men in 1952 at McCall were Ray Beasley; Howard Chadwick; David Christensen; Don Clark; Tom Dinu; Bob Donnelley; Lou Ebert; Bud Filler; Milt Goddard; Carl Holton; Ray Johnson; Gene Marker; Pete Maurek; Spencer Miller; Vern Pascal; William Payne; Carl Shaver; Charles Snyder; Art Stauber; and Bill Wallace.

The new men at Idaho City included Glenn Curtis; Jim Emerson; Charles Higgins; Ted Burgon; Elmer Huston; Gene Lewton; Roger Taynton; and Frank Tweedy.

Smokey Stover was the foreman at Idaho City and the jumpers who fought fires there returned for successive summers. Region 4 had two units, one at McCall and the sub-unit at Idaho City, both covering for each other on fires and visiting each other for weekend parties. Strong friendships were renewed at the beginning of training each summer.

We worked hard, we ate big, being young growing boys, and we thought the most important employees of the unit were the cooks. In the early fifties, they were Darlene Bell, Rocena Blackwell,

Sylvia Close, Ann Downend and Fannie White. Sylvia and Ann were the head cooks, and they made certain "their boys" never went hungry in camp. It was said in the Northwest that loggers joined an outfit for the good food, and pay was secondary. It might have been that way with us. The smokejumper kitchen was legendary. Jumpers from Missoula, off a fire and passing through the mess hall, said they were going to put in for a transfer. Some did, like Max Allen, although it was never certain that he came to McCall just for the steaks.

The jumpers had then, and still have, an unspoken bond. Each had heard the song of adventure. As the unit grew, few wanted to leave, and of those who did were drafted for Korea and later Viet Nam. Others volunteered for combat and clandestine assignments, many involving aerial operations. Others pursued careers in forestry, law, education and medicine. But many returned.

Economically, aerial fire suppression worked. The small group of individuals who started jumping in the early forties—Earle Cooley and Rufus Robinson, out of Missoula, and Frank Derry, Chet Derry and Francis Lufkin, from Winthrop, Washington—honed the skills and passed them on to their successors. But for awhile, it was trial and error as the Ford Tri-Motors and Travelaires lifted off the windy runways and dropped their adventurers into thick forests. In the forties and fifties there were few radios that would work in deep canyons, no aerial fire-retardant drops, no bladed tractors chugging up the next ridge cutting fire lanes. The work was done the old-fashioned way: with pulaskis, shovels and crosscut saws. The end result was that fires in the wilderness were suppressed early and endless tracts of timber continued to flourish in the Idaho and Montana mountains. Trees grew into maturity, died and dropped to the ground, their elements becoming part of the forest soil. Young trees sprang up in their places. There were fewer acres of black ridges, where the fires burned so hot only the mineral soil remained. Timbered watersheds continued to percolate the water to the rivers below.

Today, a few groups with strong political influence and little understanding of the forests proclaim that fire in the forest is the natural law. Do they compare the fires with hurricanes, floods, and earthquakes, which also are natural disasters? They shouldn't—man is powerless before them, but man *can* stop a forest fire. They cite the fires of the nineties as a result of doing too good a job of suppression in the decades before. These statements have been made without realizing that the forests in the Rocky Mountains had just been through a seven-year drought. Given a good start, everything in the forests could have burned, and a lot of it did. Pressures have been on the Forest Service to make controlled burns. Controlled for what? Was this the new mentality of "let's burn the forests before they catch fire"?

The vast expanses of pine, fir and spruce in the Rockies, whether near timberline or in the valleys, are not comparable to the even-aged forested stands of pine of the south, where the brush of the understory can burn off every year. Having roamed the forested stands of the Rocky Mountains for the last four decades I have yet to see a thicket of woods that I'd want to put a torch to.

We were young and dedicated, and whether we flew out from McCall, Missoula, or Winthrop, our job was to put out the fires. They were the enemy and we put them out.

Jim Larkin, piloting the Ford Tri-Motor about 1954

Chapter Three

THE PLANES

"You love a lot of things if you live around them but there isn't any woman and there isn't any horse, nor any before nor any after, that is as lovely as a great airplane, and the men who love them are faithful to them even though they leave them for others. A man has only one virginity to lose, and if there is a lovely plane he loses it to, there his heart will ever be."

Ernest Hemingway

The big silver-gray plane sat on the dirt end of the runway next to an open hanger. It was a tail-wheel plane, and its rear wheel, resting on the ground, seemed to position the large nose forward and upward to the sky, and even at rest, the plane looked as though it wanted to leave the earth and fly. The difference between this plane and the other aircraft at mid-century was that it had an engine in the nose and one on each wing. In later years professional pilots would have given a month's pay for an entry in their log books certifying that they had flown a Ford Tri-Motor.

Its wingspan was seventy-four feet from tip to tip, and the center of the wing where it joined the fuselage was so thick a man could crawl inside the wing. Its 3 engines were 365-horsepower Wright Whirlwinds, sturdy and dependable, but rarely in synchronization during flight. The fuselage was fifty feet long and could hold eleven passengers, or eight jumpers, eight firepacks, a spotter and extra cargo. The payload of the Tri-Motor was 4,000 pounds, according to the Johnson Flying Service. The cabin was so high a six-foot man could walk with ease from one end to the other. With a fuel capacity of 231 gallons, the plane could stay in the air for 5 hours.

The three radial engines, with their uncovered cylinders, swung massive propellers. These huge three-bladed props, together with the aerodynamic lift produced by the thick wings, got the plane airborne from the short grass strips in the mountains. The control cables were uncovered, outside the square fuselage. Overall, it was a stout, massive plane.

The most eye-catching feature of the big Ford was its skin. Probably never before, or ever since, has a widely used aircraft been covered with corrugated duralumin metal. The corrugations ran parallel to the air flow on both the wings and the fuselage. When the engines were out of synch, which was most of the time, the corrugated skin vibrated something awful. The noise of the three Wrights was so loud passengers had to bark commands at each other in the cabin, even moreso at the end of the summer after the side windows had been knocked out by pulaski handles that stuck out of fire packs. But the plane was beautiful and dependable, and the jumpers loved it. It became affectionately known in the aviation circles as the "smokejumpers' plane."

The Ford Tri-Motor could cruise at 100 miles per hour, and it stalled at 65 miles per hour. It could fly on two engines but not on one. If the last engine quit, the pilots would say, they'd aim the nose between two large trees, flare at fifty miles an hour to slow the plane, the trees would take off both wings and the plane would hit the ground a little slower, and the survivors could walk away. Or so the pilots said. But the jumpers weren't convinced or concerned, because they all wore parachutes.

The passengers entered the plane's cabin through a small oval door in the fuselage just aft of the wings. The door was hard for a parachutist to get out, since the jumper was clothed in a heavy canvas suit, with a coil of half-inch cotton let-down rope in one lower leg pocket, a backpack main chute, a chest reserve chute, and a wire-mesh covered leather football helmet. But he got out, at the same time the spotter was using up about half the space, crouching on the cabin floor, his head out the door, his left hand signaling the pilot to cut the engines for the jump.

The cockpit housed the pilot and the co-pilot, forward of the leading edge of the thick wing. The pilot looked past the large cylinders in the nose engine. Instrumentation in the panel, unlike the modern cockpits of today's planes, was simple: the altimeter, the rate-of-climb indicator, turn coordinator, air-speed indicator, gyroscope and magnetic compass. Fuel gauges, a tachometer and oil-pressure gauges for each engine were harder to find: they were located on the engines themselves and the pilot had to squint from the cockpit to check the readings.

Flight controls were dual, wheel-like yokes for both pilot and co-pilot. The control wheel was moved forward to drop the nose and back to raise it. Turning the wheel to left or right moved the ailerons. Trimming the horizontal stabilizer was accomplished by turning a crank above and behind the pilot's head. The crank activated a shaft that ran the length of the fuselage to the horizontal stabilizer.

The jumpers, sitting in the back, watched the pilot reach up behind him in the cockpit and turn the crank. The pilot never took his eyes off the horizon. Sitting in the back, jumpers often talked about this procedure and what effect turning that crank would have on the ability of the aircraft to stay in the air. We would find out, to our dismay, on a return flight from the backcountry. The Ford Tri-Motor had just flown over Lick Creek summit and was on a slow descent toward Payette Lake to land on the north approach at the McCall runway.

Jim Larkin was flying and looked straight ahead with his sixth sense scanning the instruments in front of him.

Gene Marker, Carl Shaver, Bob Rakozy and I, all back from a fire, sat on the floor of the cabin, listening to the monotonous drone of the engines and watching the gray peaks slide past us below the wings. We discussed the mysterious crank above the cockpit as Jim gave it several turns.

"I'm going to find out what that handle does once and for all," Gene Marker said, and slipped up behind the cockpit and quietly cranked the handle all the way to the right. Jim Larkin didn't turn around, wince, or even acknowledge that anything was wrong with the configuration of the airplane in flight.

But the nose of the Ford dropped and the plane began an ear-shattering dive with the three engines at full throttle. Not a peep from the cockpit. Jim sat there unconcerned. Gene Marker's smile faded as he grabbed the edge of the bulkhead to stay upright. The other three of us slid forward and squashed Marker against the cabin wall. Four heavy firepacks in the rear slid forward on their own, the shovel handles rattling against the open windows, until the packs came to a smashing stop against us. We grabbed anything we could to hold onto, trying to get ourselves into a sitting position to see what was going on, or at least a place to observe the crash.

We looked ahead, past Jim and over the nose of the plane. The Ford appeared in the sky like a Stuka bomber on its last dive, and Payette Lake was coming up fast. We could see the branches of the pine trees on the shore and the ripples on the surface of the water. There was neither time nor altitude to strap on a parachute, and we couldn't climb backward to the open oval door, even if we had the time. The shimmering waves were getting closer. We were losing altitude fast and getting lower—and lower—and lower. The three engines fired at maximum revolutions.

Then, slowly, the nose of the plane began to come up. It seemed like interminable minutes, but was actually only seconds, when the Ford leveled off, skimming along 100 feet over the surface of the water in the direction of the airport. Four jumpers, amid a disheveled maze of cargo packs, sat paralyzed.

"That was close," Bob Rakozy said. It was close as I'd ever seen him to getting excited.

"Man oh man," Marker said, the color returning to his face.

"We're alive!" Shaver said. "Pass me a sick-sack." He laughed nervously.

Jim Larkin turned around from his pilot's seat and addressed us in his calm manner. "You boys aren't going to do that again, are you?"

Without waiting for an answer, Jim turned back to his piloting. None of us would turn the trim crank again. I was sitting at an angle to the cockpit and I could see Jim, a big grin on his face.

Ken Roth, a smokejumper at McCall in 1946 and then a career pilot for the Johnson Flying Service, said you couldn't trim the Ford anyway. He said you had to ride the rudder or keep one wing down to fly it straight. Ken Roth and Jim Larkin were often called the best mountain pilots in the business.

The Ford with three engines also had three throttles and three mixture controls, all located in the center of the control pedestal. Another unique feature of the Ford controls was the brake, often called irreverently the Johnson Bar. The Johnson Bar was three-and-a-half feet long and was located like a floor shift between the pilot and the co-pilot seats. Pulling this big bar to the left gave the plane left brake, pushing it to the right gave right brake, and pulling it straight back applied both brakes. The only problem on landing was that the pilot only had two hands and two feet.

Jim Nylander, a friend of mine, crop-dusted in the Ford with Slim Phillips, another Johnson Flying Service pilot out of Missoula. Jim said the pilot needed three arms and three legs to land the Ford in the wind. He swears he saw Slim land once in a stiff crosswind with one leg wrapped around the Johnson Bar.

Some pilots said the Tri-Motor was a peach to fly; it could take off a runway in four plane lengths. Others said it was work to fly the Tri-Motor. It responded slow, but was very stable in cruise.

Bob Johnson, who owned the Johnson Flying Service in Missoula, bought his first Tri-Motor in 1933, and after five landings and take-offs knew it was the ship he wanted for his company's work in aerial freight operations. Later in the forties he used the plane for parachuting and fire suppression. Bob said the plane was a little heavy on the controls but "getting onto a real short strip, there was no comparing the Ford with anything else. Of the planes used at that time, the Ford would operate in half the space or less." He added, "You could turn in a 600-foot diameter circle in the Ford with a load in it. That was almost standing it on its wing tip, near vertical."

The vertical stabilizer of the Ford was big and sported the registration number, N9642. The model of the plane was 4-AT-E and it was manufactured by the Stout Metal Airplane Company, a division of the Ford Motor Company of Dearborn, Michigan. It was built in 1929 and after some time in service was purchased by the Johnson Flying Service in 1939. In addition to ferrying firefighters and dropping cargo for the Forest Service, in the forties it dropped eight smokejumpers—in two-man sticks—on fires in Idaho and Montana. Number 9642 was stationed most of the time at McCall, where it was flown by Jim Larkin, Ken Roth and Bob Fogg.

Jim Larkin flew the Ford Tri-Motor 200 hours a year on aerial jumps and cargo drops. It was a stable plane in strong winds, but once, around noon on an exceptionally gusty day, with the air moving in waves over the mountain ridges, Jim said, "We were dropping cargo over a steep ridge on a fire at Pistol Creek. Del Catlin was kicking out the cargo packs. We'd made the drop over the ridge and hit a tremendous wind gust coming up from the next canyon. The Tri-Motor went inverted against full opposite aileron control. We had cleared the ridge and the terrain dropped off sharply below us and luckily I had some altitude to work with. I glanced behind from the cockpit to see what had happened to Del. Thought maybe he went out the door. Del was upside down on the ceiling of the plane. Finally, I got the plane level again."

In 1957, after twenty-four years of continuous service in the mountains and prairies the Ford Tri-Motor—tail number 9642—and its two pilots were lost forever near Townsend, Montana. The plane was being used to spray sagebrush for the Forest Service. The Ford was loaded with 400 gallons of herbicide and while working around terrain at an altitude of 6,500 feet, the left wing hit the side of a hill, cartwheeling the aircraft, and it landed upside down 350 feet away. Pilots Penn Stohr and Bob Vallance were killed. Penn Stohr had been with the Johnson Flying Service for 25 years and had logged 12,000 hours in the air.

For small fires where only two firefighters were required, the smokejumpers used the respected single-engine Travelaire. The Travelaire, Model 6000B, registration number 9038, was an orange and black high-wing tail-wheel aircraft. The plane was built by the company bearing the same name, which was later acquired by Beechcraft. Its single Wright engine generated 300 horsepower. The Johnson Flying Service later boosted the engine to 365 horsepower.

The Travelaire, like the Ford Tri-Motor, was one of the few tail-wheel planes of that era, which when fully loaded, could still get in and out of the high mountain airstrips.

We loved the plane. After a lightning storm crossed the mountains we'd patrol in the Travelaire—two jumpers and the pilot—to look for smokes in the forests. On many a lazy summer afternoon I'd sit in the open doorway of the plane, almost half asleep, my leather boots on the outside step in the slipstream, listening to the steady drone of the Wright and smelling the exhaust smoke streaking past the door. The trees and the small mountain lakes slipped away below, the blue cloudless sky everywhere around us. In moments like this I wondered why they even paid us for the job. But then we'd see the column of smoke just over the next ridge, in thick timber and pockets of tall, gray snags, high up in the rocks, steep as a cow's face, no grassy jump spots, the wings of the plane beginning to rock from the gusts of wind, and my stomach muscles would tighten. Then I remembered why we drew a check at the end of the month.

The Travelaire was so dependable that by the end of the summer, when I was in top physical shape, I had a strange, immortal feeling that I could step out the door about anywhere and not get hurt. In a more delirious moment, looking down out of the plane at a soft green meadow, I even thought I could do it without a parachute. Larry Clark would say, "Don't worry about your chute. If you got've your logging boots on, you don't need a chute." Then he would laugh in that devilish way of his. I reminded him that *he* always wore a chute when stepping into a thousand feet of air.

Jim Larkin said the Travelaire was the best plane he ever flew. "I *lived* in that plane in the summer, logging 500 hours a year in the aircraft." The plane was equipped with skis in the winter for mail runs into the Idaho backcountry. Travelaire Number 9038 was used for many summers until one day, high above the timber north of Payette Lake, its engine stopped. Gene Crosby, smokejumper and pilot, was at the controls and he brought the plane down to a dead-stick landing on the narrow dirt road paralleling the Payette River. The touchdown was successful but the plane was not designed for driving along mountain roads and the trees soon became hindrances. The plane was torn up but Gene's skills kept it from being totally demolished. He stepped out of the wreckage unhurt. The plane was never rebuilt, but thirty years later the framework and engine were acquired by Hank Gilpin of Kalispell, Montana.

In the late forties the Forest Service purchased its own plane, a Canadian Noorduyn-Norseman. It was another tail-wheel ship, the preferred aircraft for rough strips. When landing any tail-wheel plane, the touchdown loads were absorbed by the two main wheels under the wings. The tail wheel and rudder corrections were made by the pilot to keep the aircraft straight on the runway. The Johnson Flying Service pilots were trained to "three-point" their landings with the rear wheel contacting the runway simultaneous with the main wheels. The Noordyne was a high-wing plane with an aluminum skin. Its single engine was a 1340 Pratt & Whitney Wasp, generating 600 horsepower at maximum revolutions with a constant-speed propeller. It carried four jumpers, a spotter, and cargo packs. The plane was stationed much of the time at Idaho City where the lower elevation was more suited to the plane's capabilities. The pilots called it the "lead sled" and said getting it up to an altitude of 10,000 feet was a day's work.

Another plane used briefly in the fifties was the Cunningham-Hall, a yellow biplane owned and operated by the Larkin Flying Service of Cascade, Idaho. The Cunningham-Hall was powered by a single 975 Wright Whirlwind engine, turning out 440 horsepower with a constant-speed propeller. The big, boxy plane was built for work, not looks. Its shape was reminiscent of some of the early, World-War-II German reconnaissance planes. Its fabric-covered wings blended into a fuselage of corrugated metal, the same familiar look of the Ford Tri-Motor. The entire side of the plane was door, the yawning opening perfect for loading bulky cargo, up to and including a full-grown horse. Jim Larkin was heard to say, "With the door removed for jumper work, facing 250 square feet of nothing, even the most fearless of jumpers huddled against the back wall."

Both the Tri-Motor and the Cunningham-Hall shared common heritage. Bill Stout's Tri-Motor was built by the Ford Motor Company. The Cunningham-Hall melded the talents of old-time aviation designer Thomas Hall, with the prestigious

Cunningham Car Company of Rochester, New York. Jim Larkin, who owned the plane, said, "It tried to kill me every time I climbed into it." But it could land and take off with a full cargo load on mountain strips as short as 800 feet, such as the James Ranch, where it was frequently flown for mail and supply runs.

Larkin said the Cunningham-Hall was sold to the Hood Construction Company of Boise, which wrecked it on landing at Cold Meadows. The plane was dismantled and airlifted out by helicopter, and sold to Gene Frank of Caldwell, Idaho. Gene keeps it there at Frank's Field with the Jennys, Wacos and other vintage aircraft.

The pilots might have joked from time to time about the temperaments of their planes, but all in all, the aircraft the jumpers used were wonderful. The planes were kept in excellent condition and piloted by skilled airmen. The pilots of McCall and the jumpers who disappeared out of the cabin doors had different skills, and often were separated by a decade of age, but they developed a kinship that lasted through the years.

When the wildfires got tough and the days on the line were long, we needed the pilots. After days, often weeks, smelling of sweat and smoke, our filthy black pants hanging limp from their suspenders, stomachs threatening to eat our backbones, we'd give anything for a cargo drop at the fire camp. We knew we had to work the line again in the dark, to knock down the flames that had started up from the afternoon winds, and the only food left in camp was a couple cans of beans and some used coffee grounds. Then we'd hear the distant sound of a plane, its echoes fading and growing louder along the canyon walls. We'd stop our digging and look skyward. From time to time the sound of the plane would stop altogether, along with our spirits and energies. Damn, he's going to another fire, we'd murmur together, our heads dropping back to the dusty trench in front of us. But then the plane would come around the edge of a distant ridge, and there was the unmistakable roar of the Ford's three engines. The pilot had spotted our camp in the clearing, here he'd come, through the gray smoke from the burn, just under treetop level, the cargo bags falling from the side door, their white chutes opening and flaring in the wind.

He'd make a second and third pass, dropping more cargo, each time lower than the one before. I visualized the steaks, potatoes and cans of fruit the warehouse crew had prepared. There might be a note from Reid Jackson or Del Catlin, or a radio. Maybe somebody in the parachute loft had slipped in a couple of six-packs on top of one of the insulated food containers. That would be a bonus. We'd go back to digging line and cutting trees tonight with a full stomach, and a re-supply of grub.

The pilot came around for a final pass, higher to gain some altitude. From the side door the cargo handler waved, and the pilot waggled his wings. They'd be going back to town tonight and sleep in a bed. We'd dig into the supplies and afterward go back to work on these burned-over ridges until three or four in the morning. But we respected every one of those pilots. They never left us stranded in the mountains. It was almost as though they could read our minds.

Fire Crowning
Photo courtesy of United States Forest Service

Hung Up
Photo courtesy Thad Duel

Firestorm

Photo courtesy of United States Forest Service

The Burning Mountain

Photo courtesy of United States Forest Service

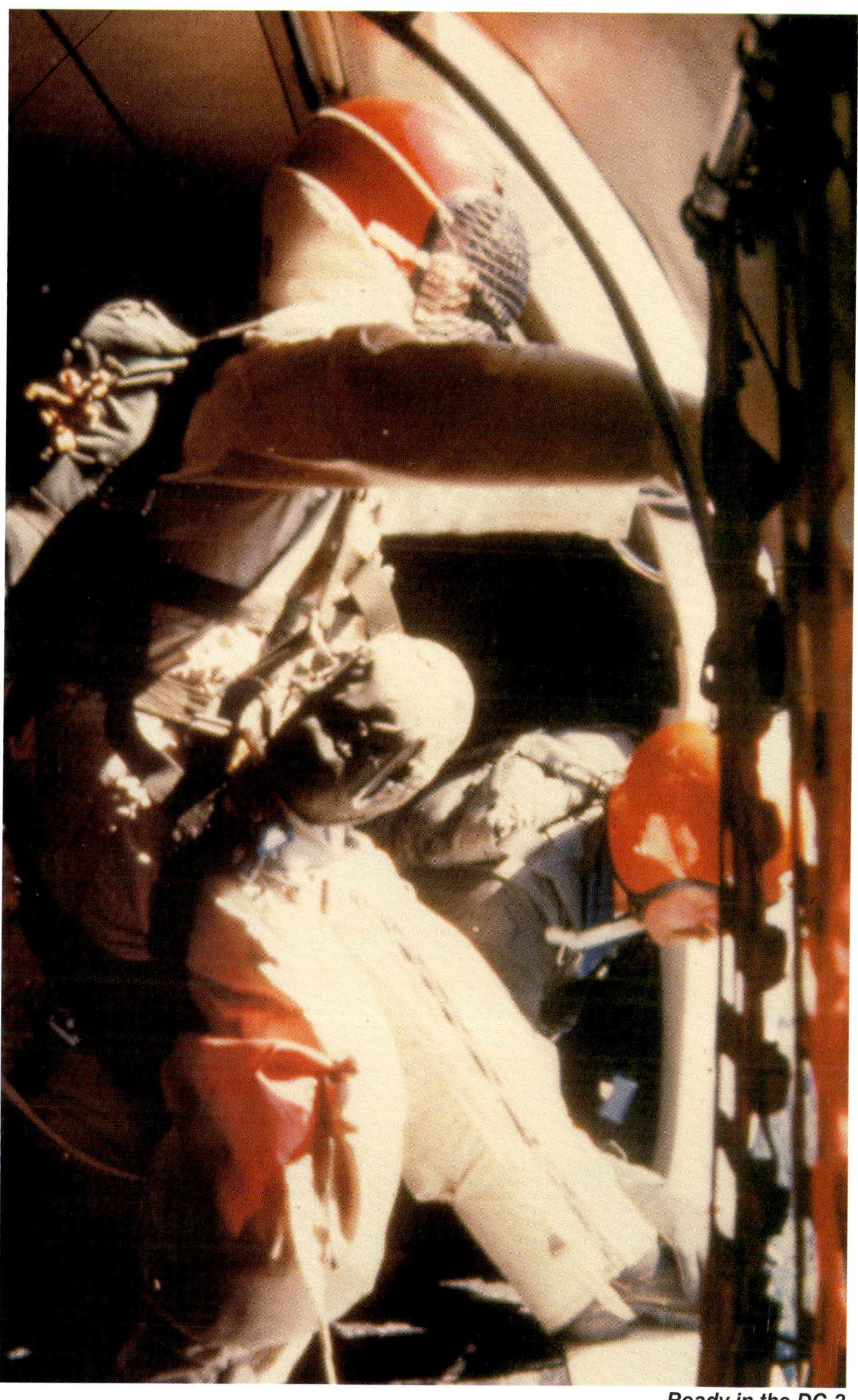

Ready in the DC-3
Photo courtesy of United States Forest Service

Reid Jackson
Photo courtesy of Reid Jackson

Left to right: Carl Shaver, Gene Ellis, Milt Goddard, Phil Hanson, Unknown, Paperlegs Peterson, Reid Jackson and Wayne Webb, early 1950s
Photo courtesy of Reid Jackson

Left to right: Bruce Yergenson, Jim McLinn, Ray Beasley, Gene Marker, Reid Jackson, Spencer Miller, Glenn Curtis, Salmon River Breaks, 1954
Photo Courtesy of Bud Filler

Packing gear off a fire
Photo courtesy of Bud Filler

Two-man Stick

Photo courtesy of United States Forest Service

Spencer Miller and the Singing Pulaski

Photo courtesy of Reid Jackson

Mel Landers & Glenn Hale, *mid 1950s*
Photo courtesy of United States Forest Service

On a clear day . . .
Photo courtesy of Thad Duel

"Get it at night . . ."
Photo courtesy of United States Forest Service

Ground Fire
Photo courtesy of United States Forest Service

Gene Crosby and Bud Filler
Photo courtesy of Bud Filler

Del Catlin

Chapter Four

THE JUMP

When the final curtain drops, the only things we truly regret are the things we didn't do.

The Sea Hunters, *Clive Cussler*

The single-engine plane flew over the fire just as the sun touched the high peaks. The storm passed after midnight and the winds calmed. White smoke from the fire Pete had reported drifted from the ridge into the canyon, and from the plane, the shattered Douglas fir was visible just down from the mountain's crest. During the night, an acre of brush and grass had burned and the black circle looked like a bull's-eye from the air. At the bottom of the burn, flames crackled in the darkness on the shadowed side of the mountain.

Carl Rosselli and I were jump partners on the call. We had jumped together on three fires and became good friends. Carl was five-feet eight and weighed in on the parachute loft scales at 150 pounds, before pulling on his heavy canvas jump suit. He had light-brown hair and blue eyes and his only outwardly Italian feature was his thin Roman nose. Carl was ten years older than I, a squad leader in the unit, and I was glad to have him as a partner. When he spoke, the words came out quick and quiet, in a way that made you want to listen. When the situation got tough on a fire, Carl stayed cool—the kind of guy you wanted to have with you on a jump.

We were ninth and tenth on the jump roster. A lightning storm had come across the Payette National Forest the evening before and the Ford Tri-Motor went out with the first eight jumpers. At nine o'clock that same evening, Del Catlin, a squad leader and second-in-command of the smokejumper unit, came into the bunkhouse. "You two guys have breakfast at four-thirty and be suited up by five-thirty." Del spoke deliberately, with authority. I dropped the paperback I was reading and Carl and I gave him our full attention. We knew it was only a matter of time before we got the call; the storm had started more than one fire.

"The Travelaire will be off the ground at oh-five-forty-five." Del dragged out the departure time, lest we forget. We wouldn't. Del was lean and tough, and kept the short military haircut. He had been a paratrooper in the South Pacific in the war and jumped on Corregidor with the 503rd Regimental Combat Team. We liked Del and nobody messed with him. "There's a fire on Pistol Creek. I'll do the spotting," he said. "Make sure Bud gets to breakfast. It'll be a long day." Del gave us his "be ready" grin.

"He hasn't missed a meal yet," Carl said.

"I know. See you two at four-thirty," Del said, closing the door.

Larry Clark looked up from writing a letter. "Does Catlin know your cargo bag is twenty-five pounds overweight with the extra grub you carry?" he asked and laughed, looking at me over the top of his reading glasses. Larry Clark, blond crewcut and blue eyes, was a high-school coach from Filer, Idaho, who walked straight and tall and was built like a professional linebacker. Larry was six years older than I and smiled a lot. I liked him from the first.

"I carry enough to last till noon," I answered. The cargo bags were limited to a three-day ration of food. Somewhere along the line I'd been unjustly accused of pushing the food limit.

Smokes reported from the lookouts were named after the nearest creek and this one would become known as the Pistol Creek fire. The Grouse Creek Guard Station reported it to the Payette National Forest Fire Dispatch office at McCall. We didn't care who'd reported the smoke. Our job was to put it out.

Del Catlin was kneeling over the open door of the Travelaire, looking for a jump spot that was near the fire but as safe as he could find in the timber and rocks. Carl Rosselli looked over Del's shoulder, the wind blowing his hair back. The only thing separating them from a thousand-foot drop to the mountains below was a two-inch canvas strap, fastened to the edges of the door. I shuffled around, moving slow in the canvas suit, trying to get a good look out the windows on the opposite side of the fuselage. This was my fourth fire jump and I was already learning to take in as much as I could of the terrain in this high, rough country.

We were studying—more from instinct than anything—the location of the fire on the slope, the fuel available to burn, and the wind. In the mountains, wind direction is usually down-canyon at night and during the morning hours, and then before noon the wind shifts direction and moves up-canyon, and blows harder. From the plane we were determining the best route to approach the fire, and most importantly, the direction to move if the fire blew up. These principles were taught to us in the classroom but you had to be there—in the air and on the ground—to make the final decisions. Carl and Del were the experienced hands at this. I was still learning.

Bob Fogg banked the plane, circled the peak and parallel to the main ridge. The engine continued its steady drone. Bob turned in his seat to watch Del at work. He was still kneeling in the open cabin door. In those years the spotter didn't wear a backpack chute. The chute would later be required by the Forest Service, after some near misses by spotters being sucked out the door by turbulence, or getting caught in the static lines. Del studied the terrain below the plane. The fire was off the ridge of the mountain, the gray smoke rolling up and over the summit. We could see flames as the brush ignited. This was not going to be an easy one to put out. The morning winds would soon get stronger.

Del turned back from the door and faced us. Over the noise of the engine, he pointed down and shouted, "There's your jump spot—the clearing below the rock slide." I wouldn't have called it a "clearing." It looked tiny but it was the only open spot within miles of the fire, about 500 yards downhill and across the slope from the smoke.

Del made a small circle in the air with his finger, asking the pilot for the first pass over the clearing, or as I preferred to think of it, a meadow. Meadow sounded softer. Bob Fogg banked the plane and brought it around once more. He kept the heading straight as he again turned in the cockpit to watch Del. The spotter was in control of the plane at this stage of the jump operation.

When the Travelaire was directly above the meadow, Del threw out a small drift chute. It caught the morning air, blossomed, and began its descent to the timber. Bob banked the plane and flew in a wide circle away from the peak.

We watched the drift chute fall. The wind was calm at this time of day, but there was a breeze, enough for me to make a mental note to avoid getting hung up in the spruces below the clearing, or in the rocks above. Going through these procedures kept my mind off thinking about stepping out of that door, and the big question: was the parachute going to open?

Part of me was saying, "Hey, don't do this crazy thing." The other part of me was saying, "I like it and there's a job to do." I told myself that few jumpers ever bit it from the parachute jump itself, just broke a leg or a back. But there was always a first time; maybe the chute packer had a hang-over the day he folded the one I've got.

Del stuck his finger in the air, asking for the first jumper. Carl, with a smile behind his mask, pointed to me. Somehow, I knew he was going to do that. On a windy day the first jumper out was a human drift chute. Carl grabbed my shoulders and pulled me around, rather hard I thought, for a final equipment check. He jerked on the canvas webbing near my waist, tugged on it for a second to see if the snaps were secure, checked the harness snap at my chest, and then the parachute riser connections and safety pins at my shoulders. Then he lifted the reserve chute to check the D-rings, and the static line. He pushed me half around again, and with that old-pro smile on his face, shouted above the noise of the plane, "See you on the ground."

This was the moment of truth: when the training, the daily exercises, the workouts on the obstacle course, the practice jumps, the mental discipline and the conditioned reflexes all came together. The question ran through my mind, as it did before, "Why do I want to do this?" Maybe I had to prove something to myself. I told myself, if I could do this, I could do anything. I thought of the story of General Eisenhower talking to a paratrooper just before he boarded the plane for the airborne drop on D-day over Normandy. "Do you like to jump, soldier?" the general asked. "No sir," the private replied, "but I like to be with men who do like to jump."

The canvas suit and the two parachutes were heavy, and I clumsily moved to the door and placed my feet on the outside step. I felt cramped in the small door opening. The safety strap across the door was unhooked and the wind slapped it back in the plane. I looked down: the gray rocks on the ridge and the trees raced along. The grass in the landing area looked soft, but I knew it wasn't. Rocks, marmot holes and old branches were hiding there. The sun was touching the ridge and I could see the high part of the snow-filled avalanche chute dropping into the darkness, which was my jump spot. Down there in that timbered hole—that's what it resembled from the air—was a piece of terrain that I'd be smacking into, one way or the other, in just a few minutes.

Carl made a final tug on the static line, which extended from the center of my chute to the longeron of the aircraft. He nodded. Right before jumping, I moved my eyes to look at the horizon, to get the best body position stepping out.

Del controlled the direction of the plane. His head and part of his shoulders were out the door, crowding me to the right, his left arm extended toward the pilot. Del's index finger pointed right, and Bob banked right, then leveled off the Travelaire. Del calculated for the wind drift, dropped his hand and the pilot came back on the throttle. The radial engine coughed once, twice, then went silent, and blue exhaust smoke streaked past my face. I felt the plane slow.

Nothing puts the hair up on your neck like hearing an engine stop when the plane is in the air, and the ominous quiet as the nose of the plane begins to respond to gravity and drop. The tail came up, the standard procedure to keep the parachute from getting tangled in the aft end of the aircraft—an unpleasant situation. I could feel the drop in the power in the pit of my stomach, which was already tight.

Actually, there was nothing else in the world I wanted to be doing at this moment. I had complete confidence in Bob's flying and in Del's spotting. Now it was up to me to concentrate on getting a good position going out the door. The slap came hard on my back: Now! The slap was never questioned. Some guys hesitated in training and they were gone. I stepped out of the door and into the blue sky of a summer morning.

My concentration focussed on the padded jump suit, past the heavy white webbing at my thighs, to the leather boots and beyond, and I was looking up—straight up—into the bright sky. My feet were coming up in front: not a good body position. My boots would soon be above my head. The plane, with the safe, secure cabin, was gone. The smell of exhaust smoke was gone; and Carl and Del and Bob were no longer with me. A jump was always solo—there was no squad leader to whisper in your ear to help you control your chute to a safe landing, no voice to tell you to pull the handle of your reserve, in those fractions of a second it took to make the decision.

Even upside-down I was aware of the sun touching the mountain peak, and the white shafts of snow below the rocks. The jump spot—the little meadow—was somewhere below me in the black timber. The only sound was the wind whistling through the mask of the helmet.

Could this moment be stopped? Could I come back to it someday? I wanted to jump a thousand times, for a thousand summers. But at the moment I was rolling upside-down, in a damned bad body position, high in the cold morning air over a remote ridge in the mountains of Idaho.

My arms were crossed over the reserve chute. I was wearing soft buckskin gloves so I could feel the risers and guidelines of the parachute when it opened, and if it didn't, I could yank on the big metal handle of the reserve. If all went according to plan, a springloaded pilot chute would deploy, catch the wind, and pull the chute from its canvas pack. If the wind was wrong, the reserve could swirl and coil around the lines of the main. Neither parachute would hold enough air and the descent would be quick—and fatal.

The combat knife was there if I needed to cut away shroud

lines that could be causing a malfunction. That's what they told us in training. But it would be hard to tell which lines. And at that point I would be falling like a rock, halfway to the ground. If I pulled the reserve chute, it would be by reflex action, a procedure seldom used but practiced over and over again in my mind. There was no fear; there was no time for fear.

I was on my back in space looking up and the parachute lines played out over my shoulder, and I could see the whirling, snapping, and popping of the white panels, now spreading, flaring, spreading again. The parachute cracked in the air and the last panel filled. My whole body snapped and there was a violent swing. Stop the swing! Stop the oscillation! Check the canopy later! I reached up for the risers and pulled them down on the forward swing. Pull and pull harder! The swinging slowed and I checked the white canopy. The panels were clear of lines. The guidelines, one on the right and one on the left, were in their correct positions.

I grabbed the right guideline, pulled it, and the parachute turned to the right. The landing spot looked small down there, past the toes of my leather boots. The forest below was still in shadows. The opening in the timber Del had selected from the plane was small, and now as I got closer, I could see that it was filled with granite boulders. Tall, dark-green timber surrounded the little clearing. On the edges were skeletons of trees deposited by an avalanche years ago. It wasn't like some of the pastures we jumped into for practice, but it was the only decent place to land. The drift of the chute took me past the clearing and I pulled on the left guideline to turn and get back into position for landing. Pull down on the risers. Plane the parachute forward and I cleared the spruces, a village of church spires as I swept over. Now to miss those boulders. The twenty-eight foot FS-1 parachute had a reputation for bringing the jumper to the ground like a bag of bricks. This morning, in the thin mountain air at an elevation of 8,000 feet, I was no exception as I was about to find out.

Keep your feet together, relax your legs, hit, rotate and roll.

The landing manuever, known as the Allen roll, takes the impact of hitting the solid earth and transfers the crash from your feet and ankles to your thighs and hips. Jumpers who landed backwards over rocks or logs could only pray their leather boots and canvas padding would keep them in one unbroken piece.

I made a final turn on the parachute to come in with the wind, on the bunch grass between two big boulders. The ground came up with a blur. There was a hit, a twist and a roll. Nothing was broken—it was a good roll. If only Rosselli could have seen that landing.I pulled an orange nylon streamer from my lower leg pocket and formed a large L on the ground, to signal to Del everything was okay. A small code detail was attached to the streamer, for a ground arrangement to signal injury, and other messages. The L meant a satisfactory landing. The loud drone of the airplane engine cut out again and I looked up to see Carl stepping out of the door into space.

A second later the crack of his chute opening reached me in the meadow. Carl worked the guidelines, cleared the spruces and turned and landed with a heavy thud fifty feet away. He pulled his orange streamer from his jump suit and formed his own L on the ground as I rolled up the shroud lines and the canopy of the main.

"Everything okay?" Carl asked.

"All okay," I answered. "It's a great morning to jump." I didn't mention my poor body position going out the door and Carl didn't give me any fuss about it, so the matter was forgotten. We were both on the ground, in two pieces, so to speak, and there was a job to do.

The roar of the plane surprised us as it came in just below the tops of the spruces. On the cargo drop, Bob Fogg approached the drop zone below the level of the treetops. Del kicked the fire packs out the door. The white cargo canopy opened, swung back and forth like a big bucket, then hit the grass with a crunch fifty feet from where we stood.

We pulled the firefighting tools from the packs as the Travelaire banked again for the last time. Del Catlin waved from the open door and the pilot did some Dutch rolls and soon the plane was a distant dot in the western sky. We stood for a moment looking at the empty sky and noticed the high jagged ridge the plane had just cleared at full power. "Up there," Carl said and tugged on my sleeve. "We had an audience on that jump."

Near the rugged summit above stood seven mountain goats, their black horns glistening in the morning sun, high on a ledge over a vertical rock wall, watching us.

Carl Rosselli

Chapter Five

THE PISTOL CREEK FIRE

Dust from the cargo drop settled around the fire packs as Carl unhooked the chute and dropped it on the grass. I pulled a shovel and pulaski from my pack. The pulaski is a top-notch fire-fighting tool, with an axe on one end of the head and a hoe on the other. It's a very effective piece of equipment, named after Ranger E. C. Pulaski who kept his crew in a cave at the point of a gun, saving their lives, when the 1910 fires blew up and destroyed a large portion of north-central Idaho.

Our fire was here and now and up in the timber above us. I dug into the cargo bag, pulled out a can of sardines and slid it into my back pocket. It was going to be a long day on that ridge and they didn't call me "Lunch" for nothing. Carl took a shovel and crosscut saw from his pack.

"Let's get our butts up there!" he said and took the first step uphill.

The sun cleared the peak and touched the meadow; the day was warming and a breeze rippled the tops of the spruces. We turned our full attention to the fire. The lazy smoke on the ridge had transformed into a nasty-looking column of yellow that moved closer to the stand of pines. Yellow and black smoke: the resins were heating and igniting, a signal that the fire was moving into the higher branches. There was more heat, more danger, and certainly more work when the flames moved into the tops. This was a bad sign, and it gave more urgency to our run across the meadow and up into the timber.

We had stowed our jumpsuits, parachutes and rigging in the clearing, the high craggy peak above us. Dense timber fell off below, and far away was the river. Just ahead the fire crackled, and smoke filtered through the branches. We couldn't see the flames but the fire was starting to move and this job wasn't going to be easy. We dog-trotted into the pines, the pulaski held tight in my right hand and the lady shovel in my left. This standard tool was a shortened version of the garden spade with the blade filed to a razor's edge. The lady shovel was effective. We'd protected the sharp edges of the tools with tied-on sections of old fire hose. The summer was at its summit and we were in good shape but the mountainside was steep and my breath came in short gasps. Carl ran hard behind me, carrying a shovel in his right hand and the rolled up six-foot crosscut saw in his left, with two fingers holding the top handle of a gallon tin of drinking water.

The wind increased and I felt it cool the perspiration under the rim of my hardhat. Above us the flames—we could see them now—swirled and growled into the wind. The brush, dried grass and dead branches were igniting. This fire was still on the ground and we had to keep it there. If the flames swept upward through the tree branches and raced to the crowns—the treetops—embers and sparks would fly with the wind, and we'd have big problems.

I scampered uphill past the clumps of beargrass, stepping over old logs. The morning workouts at camp were paying off. I panted hard, but running up and down these mountains was still easy. This was the age of strength. The world was young and exciting, and we knew we could do anything.

Carl and I moved into the resinous smoke. We could feel the radiant heat from the flames. How could we knock this fire down? Carl was the boss and we'd work together as a team, conserving our strength and using the discipline we learned at fire school. We would remove the fuel and smother the hot spots.

"Let's make a line around the bottom," Carl said. "Take the pulaski and I'll follow with the shovel." He laid the crosscut saw and the water tin on some bare rocks, protecting them if the fire beat us back to that area. I nodded and moved out with the pulaski, cutting branches in front of Carl, throwing them downhill to remove the fuel. Then I alternated with the hoe end of the pulaski, cutting the low brush that I couldn't handle with the shovel. The axe edge of the pulaski had been filed to a knife sharpness and a swift stroke easily removed the branches of the firs.

Carl shoveled dirt and built a hasty fire line as he walked. His shovel moved like a pendulum, slicing first through the grass and twigs, throwing the fuel away from the flames, and on the backswing digging a shovelful of dirt and throwing it on the fire. The loose dirt knocked the fire down better than water and the dusty soil didn't evaporate in an instant.

Carl was catching up. "Let's get an edge around the bottom and hope this wind doesn't get any stronger," he said. "We can come back and dig a wider line later. That big lightning-struck fir is smoking up there, and it's probably starting some spot fires down in the next drainage."

We put our heads and shoulders back to the line. The avenue of dirt snaked around burning logs and thickets of brush. Let them burn out—we would stop the fire below. In class we learned that a forest fire was a triangle, and the three legs of the triangle were heat, oxygen and fuel. Remove one of these legs and the fire could be suppressed. We had built 500 feet of hasty line in an hour. The downhill movement of the fire was stopped, but the wind was picking up. Both of us looked back to be sure the flames hadn't jumped the line. We were thinking the same thoughts as we turned our attention to the high edge of the burn, where the yellow smoke rolled upward.

"Go to the big fir on the ridge and check for spot fires," Carl said. "I'll keep building a line around the side and meet you up there." As he talked he threw dirt over a small pine that had just caught fire from a rush of wind.

I said okay, then moved quickly to the top of the mountain, climbing parallel to the fire. Clumps of oily salal burst into flames. It was smoky, but it smelled good. The flashes of heat quickened my steps uphill to get to the big tree that had started the whole thing. There was the lightning-struck fir, now inside the burn, and I looked up through the scorched branches for smoke. Damn it, yes, smoke swirled from the broken top. Small orange embers, little white tails behind them like miniature rockets, flew from the branches into the blue sky. This old monster is going to have to be cut down, I thought, and soon.

But first I had to check the next valley for spot fires. Carl had said to go down there and look for smokes. The squad leaders drilled that rule into us in the classroom: the main fire can be under control, but then hidden fires could flare up later and make you start all over again.

The trees and brush thinned out on the top of the ridge—less fuel. A ridgetop was the place to stop a running fire. The flames from below leapfrogged from bush to bush. I dropped the pulaski and threw dirt with the shovel, which I had picked up after leaving Carl. I needed to get higher and look into the next valley, and an outcropping of rock above might give me the elevation. Sweat ran down from the leather band of my hardhat and the salt burned my eyes. A shirtsleeve worked as a towel in the woods, until it became too dirty to use. The smell of singed hair got my attention and I looked at my forearms; got too close to the fire somewhere back there.

These were bare rocks on a sixty-degree slope, and I quickly climbed to the top, stopped, took in a couple gulps of air and looked into the next valley. There *were* two columns of white smoke down there, coming up from the heavy timber. This was going to be a race—a race against the small fires that were starting down there in the timber, a run against time and the afternoon winds that would change and come up on this side of the mountain. I picked up the shovel and the pulaski and started running. It was close, but this fire could be beaten.

Moving fast downhill was easy as I got deeper into the valley, but an empty feeling crept over me. I was alone; the ridge was back up there, and Carl was on the other side of that ridge. This side of the mountain was cool and dark, and the timber dense. But the wind here was just starting to stir. If one of these fires flared and the big winds came, this would be a bad place to get caught. It would be one hell of a sprint to the top. Back at camp we had talked about the twelve Missoula jumpers and the forest guard in Mann Gulch in Montana back in forty-nine. They never made the ridge. The fire spotted below them and, fanned by the intense winds coming up the steep slope from the Missouri River, caught them in their race to the top. Only two men reached the open rocks, running like death was right behind them, which it was. The foreman built a backfire, the wind rapidly burned an area around him, and he survived the flames.

The fires below me weren't of the size of the Mann Gulch inferno but then again you never knew.

I moved through a dense stand of old-growth white fir. The timber here, some two and three feet in diameter, was well spaced and mixed with patches of lodgepole pine.

A blue haze drifted through the firs below, the first of the smokes I had spotted from the top. I leaped my way downhill, watching the sharp edges of the shovel and pulaski, holding them a little outward. A small spot fire smoldered in the forest litter under the tree branches. I dug a trench around the smoking needles and twigs. It was easy to put out. Now to look for the second smoke. And maybe there were more.

Yellow flames flickered ahead through the trees. I picked up the tools again and ran downhill, digging the high heels of the logging boots into the loose soil, holding for a moment, then jumping, often over the fallen trunks of weathered logs.

This spot fire was hot and starting to move. The flying embers from the fire on the ridge had landed in dead branches on the ground and these flames were growing, and fast. I went to work with the shovel. Even from ten feet away the fire was scorching. The day was also warming up. Dark patches of sweat appeared on my work shirt, the sleeves already rolled up and the top buttons open. I had the can of sardines with me and Carl had the tin of water. I was getting hungry but wasn't about to eat the sardines without water.

I knocked down the flames of the second fire with more dirt, then dug a line around the burn. These spots would need to be checked again tonight or tomorrow. It was a cardinal sin in firefighting to put out the main fire and have a spot fire, off somewhere else on the mountain, flare up after everybody left.

I was deep in the dense timber and down from the main ridge, at least a half-mile on the north face of the drainage. There had been two fires and they were out. Were there more? Should I make a few sweeps across the slope and look for more smokes, or should I rejoin Carl at the top of the ridge? We *had* to drop the big fir. Carl needed help with the main fire, but Carl was cool. He would do what needed to be done. He was probably already at the top of the burn, digging the line above the fire.

Down in the dark pocket of thick timber, the wind was coming up. It rolled the high branches of the firs in changing shades of green. The wind whispered to me and it said not to linger. Years later, around the midnight campfires in the mountains, Wayne King, my hot-shot friend, told me when he was deep in the timber he often heard voices near the brooks. He said they came from the ancient people.

I decided to make two sweeps through the woods, the first one about a hundred yards below the last smoke, and the second sweep a hundred yards below that one, each a quarter of a mile laterally across the mountain. I dropped lower in elevation and made the run, watching for smokes and at the same time keeping a wary eye in the direction of the ridge top. Then I moved lower in the timber and ran the second search. Getting off course in the dark trees was easy. Spur ridges moving off right and left could sway the mental compass, but the fire itself could be easily located. Straight up in the high blue sky a thin veil of white smoke marked the top of the mountain. Carl would be digging a fireline up there somewhere.

After the search I made it back to the ridge, coming up below the rocks I had climbed earlier. Carl was nowhere in sight. I moved into the burn, my logging boots kicking up clouds of ash. Below, a silver hardhat bobbed up and down through the smoke. Carl was swinging his shovel, working uphill. For a lot of reasons I was glad to see him, knowing he was there. He joined me in a minute.

"What's the picture up there?" he asked, not even out of breath.

"The big fir is burning at the top. With this wind it'll kick out more fire. I found two smokes down in the next drainage and they're out. We'll need to check that terrain again later."

"Let's finish taking this line to the top and then cut that tree down. That snag will give us problems as long as it's standing."

"I'll go back for the crosscut," I said. We didn't ask how the other was doing. We were expected to be in shape and to work the line all day and into the night and through the next day if necessary. Complaining was a sign of weakness.

"Bring the sharpening files I left on the rock," Carl said.

When I returned with the files, saw and water tin, Carl was still digging line. My throat felt like one of the burned logs we had been throwing dirt over. We each took a long drag on the water tin, the first drink since leaving the bunkhouse at five this morning

"Let's get this big fella on the ground," Carl said, untying the thongs of the crosscut.

It sprang out flat. Carl removed the canvas guard. At four-feet thick, the trunk of the tree almost exceeded the length of the saw. Felling this monster was going to take some time. We both stepped back fifty feet to eye the tree, and I held a pulaski high to determine the straightness of the trunk. The tree had no lean, and we had no wedges. This was going to take a little luck. The smoke continued to puff from the shattered top and an occasional smoking ember bounced down through the limbs and threatened to hit us. We decided to fall the tree downhill into the burn.

Operating a crosscut saw takes some rhythm. For a cutting stroke, one sawyer pulls and his partner guides, and then the stroke is reversed. It can be a struggle to operate the saw smoothly, and four-letter words are often exchanged before an agreement is reached. The crosscut saw wasn't named the "misery whip" for nothing. But once a steady rhythm was established, the chips flew.

An hour and a half later we had almost sawn through the trunk. We were stripped down to our waists, just the red suspenders over our tanned shoulders. Our hard hats were thrown on top of the shirts. A bureaucratic fire boss somewhere would have chewed us out, if he only knew. But no one knew and we ignored the sparks dropping down through the branches.

I stepped back to check the saw cut we were making in the trunk, and looking away from our work I could see range after range of mountains to the far horizon, one following the other until they all receded into the blue distance. Maybe there was no one in all those distant hills. We were alone, sweating like lumberjacks, but I believed there couldn't be two happier souls anywhere in the world at that moment than Carl and me on that wind-swept ridge in the firs, among the dust and the smoke. It was the height of the summer. I was as far from college books as I could get, and I wondered why I would ever want to open a book again.

We had chopped out the undercut with the pulaski and the moment of truth had arrived. Would the tree fall in the direction of the undercut, or would it lean back, pinching the saw? After several more strokes, a small gap appeared on the perimeter of the uphill cut. The gap widened to a quarter-inch, then a half, and then a sound like cracking bones, like death, came from the old tree and we pulled the saw out of the cut and ran uphill. The giant tree leaned over, picked up momentum and crashed downhill into the burn, clouds of ash flying through its branches. The fire at the broken top, now receiving oxygen from a new angle, flared into life. We quickly covered the new flames with dirt.

The hard work was over. The tree was down and we'd dug a line to the bare soil around the fire. The last phase would be the search into the adjacent basin for more spot fires.

I pulled out the can of sardines. Carl produced the tin of water. "Let's take a break, then widen the line," he said.

"Sharpen yourself a stick and spear a sardine," I said, plunking myself down on a large rock.

"I see you didn't forget to bring some lunch up here." Carl smiled, wiping the dust and ash from his forehead.

"No. But I don't think this one can of sardines is what I'd call lunch." I added, "And this fire is under control."

"Let's keep an eye on it. This is the windiest part of the day and anything can happen." Carl pulled a file from his back pocket, threw his leg over an old log and began sharpening his pulaski.

This was like reloading the guns after the first battle, I thought, and, following his lead, began work on my pulaski and shovel.

We returned to the fire and widened the line to four feet. The smoking logs inside burned out. We cut branches from the small trees near the line and felt the stumps with our bare hands for hot areas, then covered them with dirt. We stirred the ashes and dirt together, and if the mixture looked fluid and hot air bubbles came to the surface we added more dirt.

"We'll check the fire again tomorrow morning. My guess is

there'll be some smokes. Then we can leave the following morning," Carl said, "after twenty-four hours without a smoke." Aside from being covered with black smudges of ash and having a woodsman's tan through all the dust and sweat on his face, he didn't look the least bit tired.

If Carl's planning was correct, we'd leave in four days, and the food we brought in would last three. Carl finished sharpening the pulaski, and picked up his shovel. He read my mind. "We'll stretch the food. Del Catlin said to pack out to the Indian Creek airstrip. The Travelaire will pick us up the day after tomorrow."

We sat on a flat rock just off the ridge. The sun dropped closer to the distant mountains. Afternoon in the high country was waning. We took turns tilting the shiny tin of water upward. The remaining smoke was thin and stayed low to the ground.

"Let's go around the fire again," Carl directed, reaching for his lady shovel.

I wanted to take a nap. "Okay, I'll grab the crosscut," I said, bowing the saw into a circle and tying the handles together with a rope. "There are three big logs to cut on the bottom of the line."

On a fire of that size, every smoke had to be completely extinguished. The dirt smothering had done the job. We removed our gloves and felt around the trunks of the burned trees and the logs.

It was time to call it a day. The up-canyon winds were dying, but air currents in the mountains were always unpredictable. Prevailing winds could come over or around a ridge, creating downdrafts, and those winds could be lethal to a fire crew. I had seen winds come up at night and the fires crown and run, often downhill. Once again, we pronounced this fire under control and angled downhill to the clearing where our parachutes and cargo packs were stowed.

"We can camp here on the edge of the meadow," Carl said, dropping his tools on the grass. "After supper we'll check the line again." It was unsaid that we would rework the fire if a flame kicked up.

Each jumper had a down military-style mummy sleeping bag, a three-day ration of food, and a small personal bag of gear. The food allotment had been weighed at the fire cache building at the base, but a growing boy often slipped an extra day's ration into the bag, or, if he could get away with it, two cans of beer, Lucky Lager being the preferred brand. Personal gear consisted of an extra pair of wool socks, a pair of undershorts, and a clean work shirt. A guy never knew when the hike out might end up in a bar in Salmon or Idaho City.

We spread the two white chutes over the grass, taking care to keep the nylon from snagging on branches. We could use the nylon as a tent if the weather unexpectedly turned wet and stormy, as it often did in the mountains. The sleeping bags were laid over the nylon.

Carl built the campfire. He shoveled down to the soil, then stacked rocks — there were plenty of them — around the edges. In the center he placed some smaller stones parallel to each other to form a small grate for heating the cans. We didn't carry cooking utensils. Cooking pans took up space and it was fun eating like the hobos we had seen when we were kids.

"There has to be some water downhill from this meadow," I said, picking up the two tins. "I'll look for a spring and be back in time for supper."

A quarter of a mile below the meadow, where the trees were dark and cool, I smelled the water. In a crevice of cold, mossy rocks a spring bubbled, the icy water flowing into a thicket of alders and reappearing below, where it began its journey to the river. I drank and drank, filled the tins, then splashed the water over my face and arms to remove the dust and dirt.

When I returned, Carl had the fire going and cans of food heating in the coals. The evening meal had been decided: canned ham, baked beans, boiled potatoes, two cans of peaches, and the standard molasses-soaked brown bread. I mixed Kool-Aid in one tin. The coffee would come last.

"I wonder what the Missoula jumpers are eating tonight?" Carl said, referring to the military C-rations they were issued in their fire packs. The Missoula jumpers envied us at the McCall unit. We could bring any kind of food we wanted—canned of course—from the base commissary as long as the weight didn't exceed ten pounds. On fires with the Missoula jumpers we would share our food. Many of the jumpers were combat veterans from the war and they had already eaten their share of military stuff. The Missoula jumpers also were issued disposable sleeping bags. On cold nights they supplemented these throw-away units by curling up in the parachutes.

The sky to the east was turning a deep blue and a cool breeze drifted across the meadow. This was the high country. We kept a wary eye in the direction of the fire for any sign of new smoke, but so far there was none.

Carl was ten years older than I, but we were more like brothers than just co-workers. There were often unspoken feelings between jump partners. I weighed 175 pounds with jump boots on, outweighing Carl by 20 pounds, but I knew he could carry me off a fire if the occasion arose, and I would do the same for him. Luckily, in the seasons we jumped together that test didn't come up.

The coffee on the fire was boiling, black and strong. Grounds had streaked over the top of the can and hissed in the fire. Sheepherder's coffee we called it. It was the best. I put on my gloves, snaked the coffee off the fire and added some cold water to settle the grounds, then leaned back against a log and lit a cigarette with a burning twig.

Carl poured the steaming coffee into two metal cups. "It was a hell of a day, but we did it. One more trip to the ridge to look for orange glows down in the next drainage, then we can turn in," he said, the firelight glistening off the metal cup.

"We may get back to base in time to be at the top of the jump roster." The dream of a busy fire season crossed my thoughts as my eyelids became heavy.

"Let's check that fire one more time before you fall asleep and burn a hole in your pants with that cigarette," Carl said, picking up a pulaski. "Take the shovel."

The winds were calm. The fire was almost out, except for the occasional puffs of smoke inside the burn. We climbed once again to the rock outcrop and looked into the next valley. For now, the fire god was beaten. But was he? He was out there, just waiting, and I knew he could return in an instant. But I had a good feeling; we had done our jobs. The timber on the mountains was green and growing. The grass and shrubs were still alive. The elk and the deer were there—I had seen their tracks. We returned to the camp in the meadow, pulled off our boots, and tucked them under the nylon chute to protect them from the morning dampness, then crawled into the sleeping bags.

I only glanced up once. Shooting stars in the moonless sky provided an eerie display of white streaks over the mountains to the north. An owl swept silently across the meadow and disappeared into the trees. Down in the dark river canyon, a coyote barked. Night fell in the mountains.

Bob Donnelley, off to a fire jump, 1955

Chapter Six

BREAKS OF THE MIDDLE FORK

The sun had not yet touched the meadow when I zipped open the sleeping bag to look at the morning. My breath came in miniature clouds in the cool air. Carl was already up brewing coffee, the old water tin perched on three rocks above the coals of the morning's fire. I pulled on the damp and smoky cotton work shirt, then resolved the next time to put it into the sleeping bag with me the night before. The shirt was decorated around the shoulder seams with small burn holes I hadn't noticed the day before. The jump boots were cold and stiff and even damper than the shirt—morning in the high country.

"What's for breakfast?" I asked, feeling guilty that he was up before me and had the fire and coffee going.

"Fried Spam, toast and orange marmalade," he said quietly, "and if you want eggs and hotcakes you have to hoof it into town."

"And that would take me until the end of summer."

It was our third day on the fire. The day before we had patrolled the line and checked for hot spots. Forest Service policy was to wait twenty-four hours after seeing the last smoke, and since the last ember was out the morning before, we could leave the burn.

We discussed how we would walk out. On the flight in, Del had said to hike downhill to the river and move downstream to where Indian Creek came into the Middlefork. A plane would pick us up on the morning of the fourth day at the grass airstrip. Carl and I figured it was about a fifteen-mile hike to Indian Creek—carrying packs that weighed 120 pounds.

"Right before we jumped I noticed a Forest Service trail on the open ridge to the south of us," I said. "A long ways off."

"I saw it," Carl said. "I think the best plan is to stay high, cross through this basin and into the next drainage, then drop straight down the open ridge. There's an alpine lake over there. We should stay above it. There's a bench there with scattered timber above the lake. We'll move past the burn on the way out, hit the high bench, and find the open ridge. Should be at Indian Creek by dark."

The word "bench" in the mountains meant a place where the terrain flattened out on a mountain slope, then dropped away again. The sides of most mountains held benches of various sizes, some so small only the deer and elk knew of them.

I knew what Carl was trying to do: keep us out of the brushy creek bottoms with these heavy packs. A guy could get stuck in the deep draws, which were filled with logs, alders and chest-high maple brush, spending hours just trying to maneuver the heavy packs a half-mile through these thickets. The draws tended to get steeper and the brush thicker the closer you got to the river.

Our gear went into a heavy canvas bag with two wide shoulder straps, sort of a giant backpack. Somebody said the bags were big enough to hold an elephant and the name stuck. I stuffed the two parachutes in the bottom of the elephant bag. Next came the harness webbing and the jump suit. The pulaski and shovel went in vertically. Then the sleeping bag, the leather helmet, and the few clean clothes, should we ever be fortunate enough to encounter a wild mountain girl on the trail. We emptied the water cans; there would be springs along the way and small creeks lower down. Carl lost the coin toss and tied the bowed crosscut saw to the top of his pack.

The packs each weighed between 110 and 120 pounds. The procedure to shoulder them was to sit down, slide the arms through the straps, then stand upright, although I had seen guys like Moose Salyer lift his pack and slip his arms under the straps on the fly.

The first steps uphill with the packs were unsteady until you got used to the weight. One thing you didn't want to do was to step on the top of a log; the high weight of the pack could flip you like a rag doll. I made that mistake just once. We grunted our way uphill, past the fire, now called the burn, and stopped on a windy saddle. Carl completed the fire report. The paperwork would be turned in back at McCall.

At noon we walked sidehill across the mountain and headed for the bench above the lake. After another hour we saw the tall spruces and glimpsed blue dots of water between them. I felt good about putting out the fire. We'd prevented this timber—majestic firs and spruces—from burning. If they had, a hundred years would go by before the woods looked this beautiful again.

Directly above the lake, rockslides dropped into the deep edge of the water. We moved to the lake like it was a magnet. A breeze played across it, and on the smooth surface, next to the overhanging branches of the firs, we saw concentric circles on the water as trout came up for the mayflies.

The bench on the high side of the lake was formed by an old glacial deposit, thousands of years before. Trees here were short, kept small by snow avalanches coming down from the peak. We moved quietly among the evergreens. It was like walking through a Christmas-tree lot at home, where I had worked for several Decembers in high school; but here the trees were living, wild and free. Several branches rolled sideways and it wasn't the wind.

"Something's moving down there. Something big and white," I said and pulled up short for Carl to get a look.

"The only thing that's big and white in this country is a mountain goat, or Bigfoot in a wedding dress."

"Well then, let's sneak up on whatever it is."

We moved quietly forward. The grass muffled our footfalls and we tried to slip up easy-like, even with the weight of the packs. A breeze rolled across the lake, up over the rocky bench, and into our faces. For a stalk on the white creature, the wind was in our favor. Whatever it was, we were getting close.

Mountain goats live their full lives in the rugged high country. They have few natural predators, except an occasional cougar, or an eagle that might swoop down and try to knock a kid from a rocky ledge. The instinct of a mountain goat is to watch the terrain below, since there are few predators up higher.

Seven goats rested on a rock outcrop above the lake, all of them looking over the water and into the canyon below. We stepped out of the trees and joined them. We could touch them, we were that close. They smelled clean, a feral sort of odor. Do these critters attack and butt? I wondered. They hadn't sensed us and we stood behind them, not ten feet away. Could I get out of this pack if they turned those black horns toward me?

Two of the little ones stared at us, the adults still oblivious. Suddenly, like we had shocked them, the little ones jumped straight into the air and landed at their mothers' sides. In unison, the five adults stood, turned and faced us. Fourteen black eyes—friendly eyes—watched us in awe. We could not easily retreat, so just stood there and looked at them, like gunfighters glaring each other down. The big billy, realizing his lair had been discovered, looked over the choices he had available. Either the goats were trapped or we were. They could go out either side, run over us, or jump over the rocky cliff. They did just that. As if on command, in the way wild creatures do, all seven turned and jumped over the edge, seemingly into oblivion.

The spell of the stare-down broken, Carl stepped forward to the edge. "Whew—I didn't think they could do that."

"I've heard they can descend anything, but look at that." I leaned over, inspecting the thirty-foot drop below.

"They did it. They jumped down on the little ledges."

"They stepped off that cliff like they knew exactly what they were doing."

They reappeared in the rocks below, moving into the shelter of the trees. "They live here — we're just visiting," Carl said. "Let's move down to the lake."

This was the fun part. The packs were heavy but at least we were going down. The shovels and pulaskis had their edges rewrapped with the old fire hose, but we were careful not to hit a branch. They could spin us like a top, and flip us to the ground. Packing more than 100 pounds in broken terrain could be dangerous work.

Mosquitoes lived in the marsh grass around the lake. We slapped the ones around our necks, but our smoky, sweaty shirts kept most of them away. At the lake, whitened ghosts of toppled logs, like narrow walkways, led into deeper water. Green lily pads wobbled in the afternoon breeze. We, like the elk, didn't stop in the wet marsh but moved quickly to the breezy ridges.

Had anyone ever been to this lake? We had found no old campfires or signs of humans. I wondered if an elk hunter had ever ventured up here in the warm days of October. Did the ancient people live here? Maybe only the spirits lived here—in the brilliant sun of the summer and in the deep winter snows. These were wonderful thoughts and they came easily. Monotonous jobs were far away. The college books were off somewhere in another world.

We passed the lake and entered a sunlit glade in the timber. The open ridge we were looking for wasn't far ahead. The deep canyon was still below us. A traveler in the mountains learns quickly to follow the ridge tops. In the draws where the springs start and form the creeks, the timber is the thickest. Mountain cherry, vine maple and alder slow the voyageur. You only cross these dark places of the forest by following the elk trails.

The packs got heavier. A drop of sweat rolled down my forehead, and I could smell my smoky workshirt. We stopped for a midafternoon drink from a spring flowing between boulders the size of trucks, the water ice cold.

I pulled a can of peaches and two candy bars from the top of my pack, as Carl opened a tin of lamb stew with his military can opener. "How can you eat cold lamb stew?" I asked him. "I'd rather eat the bark off trees."

"I thought you ate anything."

"Anything but cold lamb stew. Besides, it's probably mutton." I unwrapped a Hershey bar. "Do you eat the yellow fat on the top?"

"Sure, that's the best part, " he said, spooning out the contents. "Did you hear about the time Carl Shaver and Ray Beasley jumped on a sheepherder's campfire?" he asked. Without waiting for a reply he spun out the story. "This sheepherder was scouting ahead of his flock and was cooking mutton stew over a campfire under the trees. A lookout spotted the smoke and Shaver and Beasley jumped on it. The smoke had come from the camp, drifted up through the woods. But all was well and the sheepherder offered them some mutton. Carl and Ray wondered if it might have been his favorite ewe they eating."

"Idaho—where men are men, and sheep are nervous."

"It's true. The sheepherder had his favorite ewe. You would too, if you spent the summer in the mountains with a bunch of sheep. But the sheepherder assured them it wasn't his girlfriend."

"How did he know?" I asked, figuring Carl was setting the hook.

"The sheepherder ties a ribbon around the ear of his favorite," he said.

"Know what happens when the ewe loses the ribbon?"

"Something tells me I'm about to find out."

"He's gotta go through the whole flock again to find her."

"Sure you don't want some peaches?"

We slipped back into the packs and pushed ourselves at a steady pace down the ridge. Our shirts clung to our backs, sweat-soaked, by the time we came out to the breaks, the open country where the terrain dropped off into the river. The wind up the canyon rustled the dry clumps of grass on the hillside. We'd been losing elevation rapidly. The sun was hotter and dust rolled up behind us as we scuttled down an old game trail.

Shadows lengthened on the cliff walls along the river. Neither of us said anything but we were both searching for an easy walk under the park-like stands of ponderosa pines. Cliffs or windfalls out of sight could be tricky negotiating with the heavy packs, especially walking downhill. But we caught a break and our deer path crossed an old Forest Service trail.

"This is the one we saw from the plane," Carl said. "We can swing left and take it to the river."

The trail hadn't been used for a long time; not many people got that far into the backcountry. Windfalls covered it in many places. Elk tracks in the dust told us the big ones were nearby, and hiking became easier. We could take the switchbacks to the river, and the river to Indian Creek.

"Del told me a plane would pick us up at the airstrip in four days, unless the lookout reported more smoke," Carl said, "and that fire is dead out."

"We still have a long hike," I said.

"Should be at Indian Creek by midnight. The Travelaire

will be there in the morning."

We knew the plane would be on schedule, because we could always count on the McCall pilots to keep track of us out in the forest. A story circulated years later — it was true — about Tex Lewis and Larry Clark, jump partners on a fire above the Middle Fork. On the walk out they went up river instead of down, and took another two days to hike to a Forest Service guard station, an adventure that became known as the Lewis and Clark Expedition. Larry became my close friend and after my jumping days were over I kept in touch with him. I liked Tex, but never did jump with him. John "Tex" Lewis and Darrell Eubanks jumped together for many summers. Years later they were both killed when dropping cargo in Laos when their plane was hit by ground fire, two of the many jumpers who gave their lives defending their country in one way or another.

The elk track from the top joined an old Forest Service trail along the river. In the long evening twilight in the canyon it was comforting to see the familiar government marks on every fourth or fifth tree: a short slash above and a long slash below, blazed by a ranger long ago. It was the turn of the century before the first rangers ventured out there, afoot or on horseback, and marked the trails with a saddle axe. They followed the trails the Indians had used for hundreds, maybe thousands of years; they followed the game trails that the deer and elk had picked out along the high benches, past the springs and over the windy saddles. The mountain men had discovered the same routes when they came into the Idaho wilderness, many of them long before Meriwether Lewis and William Clark. When I saw the old blazes, now obscured with dried pitch and bark, I knew that if I'd just stay on the trail, I'd come out to civilization, sooner or later.

We stopped at the junction. The main path followed the Middle Fork, coming as close to the green water as several feet and then at times ascending to clear the rocky walls. In midsummer, the trail was dry and dusty. A lone horseman had passed a month or two before. The rounded prints of the horseshoes were faintly visible here and then there, behind a protected spot of a rock or in the soft mud where a seep saturated the path. The horseman had traveled in the direction we were moving. I wondered who he was and what he was doing in that wild country.

It was easier walking along the river. The packs were getting heavier, and we stopped more often, taking a deep breath and shifting the weight an inch or two. But the packs responded again to gravity and resumed their downward pull. I envied the horseman. He was riding. We were afoot. But we were in good shape and walking in the shadows of the evening in the canyon was peaceful, and it was good to be so close to the living world of the forest.

The splashing of the water over the rocks was cooling after the winds had beat on us on the open slopes above; the smell of mist was in the air. The trail climbed up and over the rock outcrops, and the sound of the river stayed with us. When the trail went deep into the dark firs of a side canyon the roar of the water disappeared, but we knew it was there, and when the trail turned back, the river was waiting for us.

Elk and deer prints covered the old horse tracks. A bear had also used the trail for a half mile, probably during the night. And once, we looked down at the smooth tracks of a mountain lion that had stepped out on the trail in the darkest part of a side canyon, walked the trail for 200 yards, and just as quickly left it for a thicket of mountain maple.

Immense stands of ponderosa pines filled the bottom of the canyon. Many of the trees looked to be four feet in diameter at ground level, and they were so tall and majestic the lower branches started fifty feet above the ground. These big pines, with their canopies of green protecting the grassy forest floor, created a huge park that no genius in her wildest dreams could design. This was the Idaho primitive area, a place so unknown and inaccessible that few people ever saw it.

My pack was getting heavier; we must have already covered fifteen miles. The trail smelled of dust, and puffs kicked out with each footfall. In spite of our heavy clumping on the trail, we spotted a band of elk feeding in a meadow not far above us. We stopped and watched them for a minute, until a cow issued a sharp bark, and the elk moved off into a thicket. The cow continued to bark. They would find another meadow and keep feeding.

"If I didn't know better, I'd swear there was a pack of dogs up there," I said.

"That bark came from the herd cow, and that was her alarm call," Carl said. "I hunt them every fall. If you stay in Idaho and don't go back to forestry school, we can hunt them together."

"That sounds tempting. I'll have to think about it. But if I stay in this country this fall, I may never get back to college."

Just two hours short of midnight, we arrived at an open meadow, the landing strip at Indian Creek. Carl suggested we walk to the other end and camp where the creek came into the river. "Break out the last can of beans," he said. We both sat down heavily and slid sideways out of the elephant bags.

"The Travelaire best be here tomorrow morning, or we'll have to fish Indian Creek," I said. I realized I hadn't packed any tackle, and made a vow to carry a handline and some hooks in the firepack. We ate the beans cold, and fell into the sleeping bags.

"Bob Fogg will be here, you can count on it," Carl said. I stayed awake just long enough to watch the first stars appear above the canyon walls.

By six the next morning we sat around a small campfire eating molasses bread and drinking strong coffee that tasted good in the cold, morning air. I crumpled up the blackened tin which had housed the molasses bread and which we had used to make the sheepherder's coffee. The best restaurants couldn't make coffee like this. The camp coffeemaker—he was watched carefully by the old hands—just put the grounds in the pot, boiled the water until the grounds rolled over, let them settle a moment, added a splash of cold water, then poured the steaming black liquid into the waiting tin cups.

We heard the rumble of a high-powered airplane engine far upriver, the sound bouncing back and forth along the canyon walls. Carl and I stood and turned, like a couple of mule deer that had just caught a noise on the morning air. The sunlight touched the tops of the ridges, and a trace of fog hung over the airstrip.

"It's the Travelaire, right on time," Carl said. He smiled to himself as he pulled the lady shovel from his pack and covered the campfire with dirt.

The plane came in low over the trees and the pilot cut the power, let the engine slow, and the plane dropped onto the grass. We could hear him add some throttle just before touching down.

We grabbed our packs and manhandled them out to the turn-around. The orange and black Travelaire rolled up and rotated 180 degrees in a cloud of dust, facing back down the field. The big prop shook once and stopped. The side door opened and Bob Fogg, crouching low, stepped out.

"Morning. Doughnuts and coffee in the front seat. Saddle up and let's get moving. The forests of Idaho are burning down from that last storm. You'll probably be back out before the day is over."

That was a lot of conversation for Bob Fogg. Bob was serious, quiet, a consummate professional pilot, considered one of the best mountain pilots of the Johnson Flying Service. He stepped away from the plane and lit a cigarette as we stowed the fire packs.

He had a policy of flying in and out of the back country strips only in the morning hours or late in the evening when the air was cool and the winds were calm. In the middle of the day when the air was hot, the pilots stayed away from the mountain strips.

Flying could be disastrous to aircraft and pilots attempting to land in those mountains, but the airmen of the Johnson Flying Service were among the best. Bob Johnson, who started the company back in 1929, required his pilots to log at least a thousand flying hours before they could land into a backcountry field. This morning, as usual, Fogg wanted to waste no time getting back into the air.

There were others who took chances, and the scattered wreckage of planes could be seen around many of the mountain fields, the red or white crumpled wings slowly being reclaimed by pine needles and huckleberry brush.

Bob gunned the engine, the Travelaire rolled out easy, the nose came up and the plane cleared the pines by a hundred feet. This was a perfect morning to fly with the blue sky above and the green canyon walls gradually getting small behind us. The plane banked to the right and he took a heading to the small town of McCall, seventy-five miles to the west, over some of the highest and most remote peaks on the North American continent.

❖ ❖ ❖

McCall, Idaho, is nestled on the southern shore of Payette Lake. In 1952, it was the home of a few hundred loggers, sawmill workers, Forest Service personnel, storekeepers and bartenders — mostly bartenders. The population in the summer doubled as the well-to-do from Boise traveled the 100 miles north to escape the southern-Idaho heat by waterskiing and fishing. The town itself lay in a long valley filled with cattle ranches and surrounded by timber on the hillsides.

Our plane made a descent over Payette Lake, came in above the town, and the tires kicked up dust as they touched down on the dirt. After a short taxi, we stopped next to a brown building. Bob turned the airplane into the wind and cut the power.

A Bell helicopter had just landed on the asphalt next to a hanger with a sign above the door that read JOHNSON FLYING SERVICE. The Bell, a small ship, had been introduced that year for fire spotting and medical evacuation. The door opened and the pilot, tall and suntanned in starched khakis, stepped over the left skid. His passenger, a good-looking man in his late twenties, with athletic shoulders and biceps that looked as big as footballs, tucked his tan shirt behind his belt, dropped his head to clear the drooping helicopter blade, and deftly stepped over the skid.

This second occupant was Reid Jackson, foreman of the Forest Service Region Four smokejumper unit. Reid had wavy sandy hair, stood just shy of six feet, and had the look of strength about him, like a weightlifter. He tended to move quietly and quickly, with his head and shoulders slightly forward like a cat, or like a person in a hurry to get the job done. Reid had a degree in forestry from Utah State College, and in addition to other forestry jobs after graduation, he had worked his way up through the smokejumpers to his job as foreman. His first love was being out on the line with the other jumpers.

Reid was a first-class spotter and when the camp was empty, and it often was, he would strap on a parachute, find a jump partner, and fly out to drop on the next fire. It was a small, close unit. We knew him well and respected his decisions.

I was unloading the packs from the plane when Reid approached. "Welcome back," he said. "Catlin told me you were going to have a long walk." He took in the appearance of our equipment and at the same time assessed how much stamina we had left.

"You have the honor of being the only jumpers in camp," Reid said. "Take a shower and repack your gear. Sylvia and the cooks have been waiting for you. They'll fix you up some steak and eggs."

Sylvia, the head cook, was hard-working and conscientious. She'd wait in the cook shack most of the night to feed the boys—her boys—when they returned, tired and hungry, from a fire. Formal meal hours were cast aside when the fire season was rolling. Sylvia and her assistants kept the kitchen warm and cozy, smelling of baking bread and blueberry pies. The steak and eggs would be just the beginning. On the white tablecloths there would be orange juice, fresh peaches, toast, butter, fried potatoes, hot cakes, maple syrup, and pitchers of cold milk. After eating out of cans for four days, I had visions of food just thinking about it.

Reid helped us throw the firepacks into the back of the pickup. "Pat Daly and Bob Donnelley jumped on a fire on Fall Creek yesterday," he said. "They couldn't hold it, and we called in eight jumpers from Missoula. They jumped two four-man sticks. One of the guys hit a gust of wind and came in sideways and broke his upper leg on landing. It took four jumpers to make a stretcher and carry him to a clearing on a ridge. The pilot and I just came back from there. These choppers don't have much power over 6,000 feet, and that ridge was at 7,000. We strapped the jumper to the right skid. I stayed on the ground and the pilot flew him to the hospital at Cascade, then came back to pick me up. In the meantime the fire kept moving. Nine jumpers are on it and a ten-man ground crew is walking in from Landmark." Reid paused and Carl and I climbed into the back of the truck. He pulled himself up with us. He could have sat up front.

A broken upper leg: the femur, largest bone in the human body. The first-aid instructors told us a person could die from a broken femur—die from shock and internal hemorrhaging. First aid was to apply traction on the leg and hope an artery wasn't ruptured inside the thigh muscles. Two poles could be cut, one placed under the armpit and the other in the crotch, padded with the jump jacket, and traction applied by twisting a cravat or T-shirt tied securely to the boot. I wondered how it happened. Must have been more than a gust of wind. Maybe he hit sideways against a log or a boulder. The details would come out when Pat Daly and Bob Donnelley returned.

The pickup turned right on the dirt road to the jump base. Reid sat with his back to the cab. "We have our hands full," he said. "There are still some fires unmanned. After you have breakfast, repack your gear, check out new chutes from the parachute loft, and be ready to jump. Bob Fogg will fly the Travelaire back out, and take you over the breaks of the main Salmon. The lookouts reported two smokes going strong out there."

It was nine in the morning, and the day was just beginning. Carl and I didn't know it then but two full days would pass before we'd sleep again. Now as I listened to Reid I felt an extra thrill in knowing we were the only jumpers left in camp to stop the fires. It was like the movie I once saw of the last two Marines on Wake Island manning a machine gun as the Japanese came through the surf and up the sand.

Let's go back to the sky, back to the mountains. The possibility of having a couple of beers that night at the Foresters Club was cancelled. This was what I had signed on for, and I loved it.

1950 smokejumpers outside the ready room

Chapter Seven

HOLDING THE LYNX CREEK LINE

"Never give up, never, ever give up."

Winston Churchill

Carl and I prepared new fire packs in the fire cache building, the small warehouse across the yard from the parachute loft. We grabbed sharpened pulaskis and shovels from their storage racks, and strapped protective canvas guards over the cutting edges. A three-day ration of food, this time including more cans of juice, was added to each pack. The sleeping bags from the last trip were dusty but in good condition and we stuffed them into the bottom of the elephant bags to cushion the cargo drop. Two number-ten tin cans of water were placed on the tops of the packs, to keep them from rupturing when they hit the ground on the cargo drop. I carefully tucked the coffee and a pack of Luckies near the top, for first out, although Carl didn't smoke and I smoked very little. But a guy could go berserk in the mountains without these essentials. The story around camp was that Gene Ellis, a squad leader and a tough boxer from Idaho State, once forgot his coffee and cigarettes. His jump partner said he figured he might have to kill Ellis before they finally put the fire out. They walked ten miles to a fire tower where, luckily for the lookout operator, Ellis bartered for the necessary items.

After a late breakfast at the cookhouse, where Sylvia fed us like kings, we cleaned up and put on fresh work clothes, laundered by hand—our hands— at the Forest Service washing machine behind the bunkhouse. We lacked the skill of our mothers, so the work shirts came out wrinkled and the white T-shirts had a blue tint.

Every profession evolves its own clothing, and parachuting and forest firefighting was no exception. The most important gear was the footwear. Most jumpers wore the thick-soled all-leather high-heeled logging boots manufactured by the White Shoe Company of Spokane, Washington. The boot was heavy but the weight of the body was directed to the center of gravity at the heel. The foot was firmly enclosed in stiff leather and you could leap from log to log and up and down steep slopes all day without tiring. Many a jumper had come down through a dead snag, snapping off branches like shotgun blasts, landing on the ground next to the weathered trunk, without so much as a sprained ankle. The cost of a pair of Whites in 1952 was $24.50, compared with $250 today. Some of the men wore Bone Drys or Curran Greens, but the preference was the Whites with their all-leather last. A pair of boots could make it through the summer unless there were days spent mopping up in hot embers in the burn.

The pants were the faded black Frisco jeans, loose fitting and cut short at the boot tops, like the loggers did, to prevent snagging on branches. Suspenders held by buttons allowed for fast and easy movement over logs. Shirts were long-sleeved blue or tan cotton, or the hickory-striped shirt of the lumberjack. The flat metal British-looking hard hats were worn, but only on fires where snags or burning branches were dropping. Most of us left them in our fire packs. Later regulations required wearing them on any type of fire. It was a good idea.

The laminated aluminum fire tents had not been developed in those early years, and if it had been left up to the individual jumpers, they probably would not have carried the tents. They could give the firefighter a false sense of security. The first thing to use to prevent burning up in the woods was the human brain. Were the fire tents successful? The government officials seemed to think so. The hotshots and jumpers who would be roasted in them in later years did not testify.

Soft buckskin gloves were used for jumping and working the line. The gloves prevented serious rope burns on the descent, and on the line they protected the hands from blisters after hours of swinging a pulaski. Since the buckskins were expensive some of the guys just folded them after landing and stuffed them in their helmet, working the fire without gloves.

To pick up the new parachutes, Carl and I walked across the dirt driveway to the loft. The riggers were out on fires and the selection was limited. Most of the chutes were out or hanging in the loft unpacked.

"Here's Number 102," Carl said, fussing over the last of the packs.

"You can have it," I said. Number 102 was one of the original FS-1 parachutes that had a record of malfunctioning. Some parachutes were infamous for not opening like they were designed, and old 102 had a poor record. It deployed inverted, it deployed with line-overs, and sometimes it didn't deploy at all. The jumper had some anxious moments before he pulled his reserve.

We tossed a coin. Carl lost and got Number 102. The repartee was repeated in selecting the reserves. It would be the last time the four chutes would be used.

Reid pulled up to the loft in the pickup. "Slim Vassar just called. The two big fires are still going strong out there, and the lookouts in the last fifteen minutes have reported two new smokes. These last two are probably sleepers from that storm." Bob Fogg was right — Idaho was burning down.

"The fires are east of Chamberlain Basin near the breaks," Reid said, grabbing a firepack and throwing it into the back of the pickup, like it was as light as a feather. "Do the best you can until we get more jumpers. You'll have to spot yourselves. Bob Fogg is flying and he'll report back on the fire you choose. Watch the winds. It's hot and gusty this afternoon." Reid swung into the cab, and we climbed into the back.

Reid started the pickup and dropped the shift into low, as if there were an urgency to this trip. We had already pulled on the canvas suits with the high collars. The truck kicked up dust until we turned onto the asphalt. The blue water on the left almost touched Lake Street, allowing just enough room for the bars and the boat docks. As the truck picked up speed, three people waved from the service station on the right. Nearby was the Foresters Club, smelling of stale beer and cigarette smoke, where just last week a cute redhead on the next barstool had smiled at me. I made a mental note to go back to the Foresters after this fire was out.

The tops of the pines, just over the Hotel McCall, were swaying, and whitecaps ran from Brown's millpond to the west shore. The take-off would be crosswind, but tougher than that, the winds in the high country would be howling. Riding to the airport through town in the back of the pickup with Carl, past the townspeople, made me feel proud that I could do this job. There was a purpose to jumping on these fires. We were going out to a fire that could destroy thousands of acres of trees and devastate a wilderness forest for decades. *I hate to see a baby tree burn;* the dialogue from the movie went through my mind. Men have jumped from airplanes for years, but did they jump in the mountains, in the rocks, the timber, and in the afternoon winds? I had proved to myself that I could jump, I had shown myself that I could run, work and survive in the mountains with Carl, the squad leaders and the other jumpers.

An inner strength tied us together. I think we all felt the same about the unit and about each other. We rarely expressed these thoughts of our brotherhood, not in the mountains around dying campfires, nor in bars in town, but the bonds were there. I knew. I could tell, and years later, when I left and would meet an old jumper friend on the street or in a watering hole, those unwritten codes would come back. I'd carry you off the mountain if I had to, and you'd do the same for me, and I'd back you in any barroom fight, even if there were only the two of us, and you'd do the same for me. There was something special about that job, even a little heroic. More people waved and I waved back but I was also thinking of those whitecaps just off the boat docks.

The Travelaire was tied down next to the Johnson Flying Service hanger, its nose facing into the wind. The side door stood wide open, waiting for us. Next to the wing, Bob Fogg stood smoking a cigarette. We said our greetings, then climbed stiff-legged into the plane. Our lower harness snaps were unbuckled, allowing us to move around easier in the plane. We'd make a final check of the snaps before the jump.

"We're flying northeast to the main Salmon," Bob said, grinding the cigarette out in the dirt at the edge of the taxiway. The word "main" was used to differentiate the Salmon River from its Middlefork and the Southfork branches. "Dispatch reports some big fires out there. You guys are the only ones left to stop them, until ground crews are relieved from other fires. Let's go up and see what they look like." Bob slid into the left seat and Carl settled on the cabin floor. I fastened the safety strap across the open door, then sat next to the firepacks. The engine choked, roared and gained revolutions. Exhaust smoke flew along the fuselage. The noise was too loud for Carl and me to talk but our eyes met for just an instant, and I knew we both felt the same way, the high adrenaline of the adventure: flying off to an unknown fire on this windy Sunday afternoon in July.

Dust blew from the sagebrush across the runway as the plane lifted off with its right wing dropped into the wind. At 500 feet above the ground the plane banked to the east. Ahead, Jughandle Mountain loomed, and a chain of lakes lay shimmered off the left wing: Boulder, Anderson, Maloney, and Buckhorn. Soon the canyon of the South Fork of the Salmon was below us, and beyond, the rolling mountains surrounding Chamberlain Basin, part of the large primitive area in central Idaho. To the south we could see fires that had been burning for days, by the look of the smoky haze in the valleys. Most of the fires were manned by jumpers and ground crews.

The sky in the west behind the tail of the plane was filling with high gray clouds, not the towering thunderhead type, but ominous-looking anyway. Shafts of rain dropped from the clouds to the higher peaks. The rain probably evaporated before it touched the ground. The mountains were too dry to accept much moisture. The weather was changing, and not for the better.

An hour later we closed on the canyon of the Salmon. Steep forested mountains, shiny lakes and bright meadows passed below us. The hum and vibration of the plane's engine, the smell of oil, and the warm wind coming in the side windows was comforting, and I found myself relaxing. The engine of the Travelaire droned on for hours without missing a beat. There was a sense of power and dependability to this plane, a feeling almost euphoric, being there in the cabin with two friends, the big sky above, and the green carpets of timber stretching for miles below.

Then we saw them. At a distance of twenty miles, two plumes of white smoke broke the background of the mountains. Bob changed the heading to the left to check out the first fire, the larger of the two and it came up fast under the nose. The second smoke looked about ten miles off, back from the canyon rim on a long slope of grass and scattered timber. We banked and circled the largest of the fires.

This blaze already covered more than twenty acres. The fire was on the rim of the forest at the top of the canyon breaks. We lost some altitude and flew directly across the fire and straight over the Salmon River, shining in the sun. The canyon dropped off fast below the plane. A fire had swept across the rim of the canyon years ago and the rolling hills beyond were covered with thickets of doghair lodgepole pine, the first species to regenerate after a fire. The trouble with this forest was that most of the burned snags from the old fire, now whitened and weathered, had fallen and were criss-crossed for miles below the new trees. Val Simpson, the ranger, had said a person could walk on logs from one side of Chamberlain Basin to the other without stepping on the ground—an exaggeration perhaps, but this terrain was no easy place to fight a fire. To dig a line in that timber, the deadfalls had to be cut with the misery whip, a serious challenge for two guys.

Bob banked the plane again and made another pass over the fire. We watched the wind direction, eyeing the ground for a jump spot and looking for a rendezvous point and fire camp. The fire was on the move, and pushed by the afternoon winds it had consumed some timber and was going to kill more. Individual trees ignited, crowning to the top like giant Roman candles, as if to gather strength, build up more heat and run across the rolling hills. There was danger in the afternoon in that beautiful canyon.

"The winds are too strong, too gusty to jump," Carl shouted above the roar of the engine. It was the windiest part of the day; the gusts were strong and blew in all directions. That high gray sky was getting closer. The descent of a parachute could be sideways, or even back up, if we got caught in a thermal, an invisible mass of air coming up the side of the mountain.

Bob turned in the front seat. "There's no shortage of fires for you guys. There are two more small smokes on the other side of the canyon." We had also seen them, the ones Reid said had just been reported. "Let's go back and land at Chamberlain, wait until this evening when the winds are calmer, and then you can decide which fire you want to tackle."

Carl nodded. We flew over one more time and Bob leveled the wings of the plane for the flight to the Chamberlain airstrip. It was rare when jumpers didn't go out the door in the afternoon over a fire, but the winds were potentially deadly. I looked back at the smokes, now getting smaller in the background. I was anxious to get at them.

A Forest Service guard station was located at Chamberlain Basin. Landing into the wind on the grass runway was easy, and Bob taxied the plane to the log cabin at the edge of the trees. The propeller flashed in the sun, then stopped. We stepped down from the plane, stiff-legged from sitting in the heavy gear. I unhooked both parachutes and dropped them on the ground. It was time to take a nap.

"You think you're tough enough to go back out this evening?" Carl asked, smiling, and noticing I was looking for a place to fall asleep.

"I'll be ready."

"We'll wait here until seven-thirty, for these winds to die." Carl took off his jump suit and I followed. The three of us walked to the guard station in the shadow of the pines. Bob knocked on the door and then opened it. The cabin was empty — the Forest Service guard and his crew were gone, probably out on a fire. We went in and moved carefully, not wanting to disturb the well-kept kitchen and the living quarters. There was an unwritten code of ethics in the backcountry: the owner left the door unlocked and the traveler of the woods could help herself to coffee, sourdough and a can of beans. The traveler was expected to clean up, and split some firewood before departing.

The warm afternoon in the meadow seemed endless. We

dozed in the grass in the shade of the forest's edge. Bob Fogg had found some coffee in the guard station and was enjoying a cup and a smoke on the front porch. Insects hummed in the meadow. Large summer flies, in squadrons of two and threes, departed the porch of the cabin and buzzed our heads. The tail-wheel plane was parked quartering to us fifty feet away, and heat waves shimmered over the wings. I looked up. The plane appeared to be watching us, watching and waiting. Like a Labrador retriever, it was ready to jump when it got the word.

I looked to the sky, checked the weather. Long cloud plumes streamed off the peaks, indicating strong air currents at the higher elevations. When evening came, which fire would we attack? The two small smokes across the canyon looked easy. Jump on them, put out the fires, and pack out. The biggest challenge was the big one on the edge of the canyon, which had probably doubled in size. Hard to tell, but when we flew over the fire earlier, it looked like a twenty-man operation.

Carl and I discussed the strategy and decided to jump on the largest fire. It was the most treacherous and could take out the most timber, and could burn until the end of the summer if not stopped. We could at least try to hold it from moving downhill into the canyon. If we could hold the bottom and maybe even the lower flanks, the crews coming in the next day could tackle the front. Trying to stop the lodgepole from crowning at the head of the fire with two guys was almost impossible, if not downright dangerous.

Interesting word, I thought. We rarely called anything "dangerous." It was fire knowledge, training, and escape tactics we talked about and worked with. The big fire would be a hell of a problem for two firefighters, but like any big problem, we'd break it down to its elements, and tackle them one at a time. We'd work the lower part, which was burning its way downhill. Because of the steep pitch of the slope it wouldn't run fast. But it could get damn hot there. And that is where we would be.

"Let's saddle up. It's seven-thirty," Carl said, and snapped up his jump jacket. Bob completed the ground check of the plane and climbed into the front seat. I pulled up the jump pants and had that strange feeling something was watching me. It was the plane again; that plane had smiled at me. Like a spring-loaded dog, it was ready for action. But we were the participants on this windy evening. He — the plane, the Travelaire — could stay high in the sky and watch. Just a hallucination, I thought. I had taken too long a nap.

Part of the runway was in the shadows as we lifted off and turned northeast to the darkening canyon of the Salmon. In twenty minutes we circled the largest fire at an altitude of 1,000 feet. The fire had grown another five or ten acres during the afternoon, into the jackstrawed lodgepole. The wind gusted, wanting to change direction. Regardless, we were going to jump. At least it was cooler now.

"There's our spot," Carl shouted, leaning out the door. "Right on the rim next to the fire." It wasn't a spot, but more like an opening at the top of the ridge, inconveniently filled with short, scattered whitebark pine. It looked easy: not much grass, and I could see stretches of bare soil. Not many boulders either. Just the unrelenting wind. *Don't think this is too easy and don't relax.* Thinking a jump was simple could get a guy in trouble.

We were spotting ourselves. "Drop the cargo bags 200 yards from the edge of the fire. We can camp right there," Carl yelled, pointing out the door to the clearing. Bob descended to 200 feet, sounded the buzzer in the cockpit, and I kicked the cargo pack out the door. It landed seconds later in the middle of the clearing.

With the plane at full power we went back up to an altitude of 1,000 feet above the open ridge. I threw out the drift chute to check the wind direction and we watched the small white canopy drop into the lodgepole. We already knew what the wind was doing by watching the smoke. It was gusty but we were going to go. Carl gave me a thumbs-up sign.

Bob banked the plane for the jump pass. I sat in the open doorway, feet on the outside step, the prop wash hissing through my facemask. Daylight was fading. Orange flames flared below as a tree ignited. It was fun spotting yourself; there was no time to think about the actual jump. I gave a hand signal to the pilot, and the radial engine coughed and stopped. The nose began to drop and I stepped out into the dimming light, listening to the fast-fading hum of the air past the struts.

The parachute opened as it was supposed to, but the shock

was like being snapped at the end of a giant whip. I swung back to a vertical position and pulled on the right guideline to face back into the top of the canyon. Now to gain some distance upwind from the spot on the ridge; I could always change direction and fly with the air. Turn too soon and I'd fly past the landing zone and hit in the deadfalls or the fire. Neither would be fun, and I could wind up with a broken leg. At 300 feet above the ground the big parachute turned. The whitebark pines that had looked small from the air were larger now and coming up fast. I was low and slow, as the pilots say, so I held the risers down and planed into the open hillside. It worked, and I came down in the middle of the clearing. The fire was only 100 yards away — too close for comfort.

The sky was getting dark. Much later and a jump would be too hazardous. The engine of the plane stopped again and Carl was in the air. He got the feel of the wind quickly and planed away from the rim, then closer to the ground than I would have done he turned, pulled the risers down to his chest and came back into the open clearing. He said later he used my white parachute as a guide in the fading light. Twenty-five feet from me he hit and performed a perfect roll, as only his wiry frame could do, and in a cloud of dust he jumped up to a standing position. He said not a word, just a big grin on his face as he unsnapped his reserve and dropped the rigging to the ground. We put out the orange streamers, then watched the fire off to the side, the yellow flames cracking and rumbling. Bob Fogg, circling above, acknowledged our markers, rocked the wings once and turned in the direction of McCall. It would be pitch black when he got back to the airport.

Darkness was upon us, and the only sound was the whisper of the wind in the trees. For a brief moment I felt completely isolated, maybe even lonely. The breeze across the ridge ruffled my shirt collar and gave me a goosebump or two. I glanced at Carl, busy gathering up his jump gear. Nothing bothered him. We had a job to do, and the feeling passed quickly.

"Old Number 102 opened like a champ," I said.

"It worked like it was supposed to," Carl said. "Maybe those other guys did somersaults in the shroud lines."

We stopped at the firepacks and cargo chute. I took a long swig from the water tin to get rid of the dry taste in my mouth.

Jumping did that.

"Let's move the packs and the gear over there, and make a dry camp. We'll work the fire like we said, down the lower flank, across the bottom, and up the other side. I think we can do that and be back here by midnight. The parachutes and packs should be safe."

Without a word — just like the Pistol Creek fire — I pulled the pulaski from my pack and Carl slid out his shovel. We removed the guards and faced the fire. It was growing larger, crackling into the pines away from the rim of the canyon. We could do the most good on the bottom.

We started the line, using the natural openings of the hillside at every chance. I cut alder and cherry brush and Carl threw dirt. The gray clouds I had noticed earlier hovered above us and the night was black as coal. The glare of the flames became hypnotic. The burning and crackling of the needles in the branches taunted us. *Try and stop us*, the flames said. *Come closer, closer, and we will kill you*. We threw dirt back. The flames died and we moved on. The wind continued to gust, fierce at times and coming from different directions.

We had dug line down one side and started across the bottom. Flaming cones were rolling downhill. We stopped them with small trenches, and moved on to the next fire. Snags or logs uphill could tumble or roll. When we had the chance, we worked the line across openings, away from the falling perimeters of old trees. These were tricks not taught in the classroom, but learned on the line. We constantly watched the fire. When you couldn't see it because of smoke or terrain you got into trouble, and it didn't take long.

The winds increased. Instead of cooling, the night air got warmer and swirled around the rim of the canyon. A finger-like ribbon of fire, fanned by the winds, had moved into a pocket of Douglas firs. The brush below the trees ignited and the needles of the firs became superheated. Then the first tree exploded, the resinous needles flaming like gasoline and roaring to the top. A heat wave pushed us back fifty feet, the radiant flash of fire against our faces.

We watched the flaming twigs and smoke disappear into the night sky. Another fir ignited, then another, in stark yellow flames. There was nothing we could do but back away and drop lower.

Our only option was to cut paths through the brush thickets and pull small logs away from the fire. Dirt from Carl's shovel collapsed the flames on the lower branches, and we watched the dying embers swirl away into the night.

The clouds moved on and a crescent moon came up over the mountains across the canyon. It was one in the morning and we stopped for our first break.

"We're slowing the fire on this edge," Carl said, wiping his forehead with a red handkerchief. He was always optimistic.

I wished we had a drink. We hadn't brought water from the packs and got used to going for long spells without drinking. It was easier that way. I felt damned tired: four days on the Pistol Creek fire, and now working through the night on this one, whatever it was called. The water and the food were up there somewhere on the ridge.

Carl tossed the shovel into the dirt, like a jackknife. The handle quivered. "These winds keep taking this fire lower and lower. It could have blown up and run on the top. Can't tell from here. Couldn't do anything about it if it did, just the two of us. We'll do the best we can down here. It'll be daylight before we dig this line around the bottom and back to the rim." Carl had smudges of ash across his face, and I noticed from the glare of the flames his shirt was streaked with coal-black smudges.

"We can walk over the burn, back to our packs," he said. "Then we can decide where to go after that." Somehow, sleep was never mentioned. Carl was looking uphill at the fire, and from the look on his face, he wasn't even considering giving up the fight.

On the other flank of the fire, a tall fir ignited and the flames raced to the crown, consuming the tree in twenty seconds. A second tree crowned in an orange ball, and again, a third.

Carl grabbed the handle of the shovel like a baseball bat and started uphill. "Let's stay way below those trees and work the line through the grass on the left flank."

"Watch that burning log below the firs," I said. "It just might roll." This was no time to relax.

The moon was high, and a faint glow of dawn silhouetted the distant peaks. I almost felt like lying down right there on the dusty hillside and going to sleep, and the hell with the fire. It was getting tougher swinging the pulaski and my mouth was so dry it felt like I had been eating cactus. The flames reflected in Carl's eyes and off his metal hardhat. He didn't even look winded.

"Thought we'd find a spring along here somewhere, where a guy could get a drink," I said. "Never do carry a canteen because there is usually water in these mountains." I didn't dare stop; it would be too hard to get started again. There was a moment in the middle of the night when the pulaski started to get heavier — and heavier. My eyelids were that way too, but I fought the feeling. With the dawn it would be better.

"There's a small creek over the top and just down the other side, about where the head of the fire is. I noticed it from the plane. We've also got the water in the packs," Carl said.

Water and three cans of fruit would taste good right now. I turned away from the mountain and saw the morning sky to the east, and the thought of a cold spring up on the top made me aware of how tired I was. Didn't Carl *ever* slow down? Perish the thought. There was more fire here than a dozen jumpers could handle. Just keep moving, I told myself.

The pulaski cut a path in the dirt about as fast as a guy could walk. The short shovel in the hands of an experienced firefighter could smother the flames better than a fire hose. We pressed on.

The trees on the backside of the rim were crowning, flaring from the bottom to the top, but had not gathered momentum yet for a blow-up, which might happen if the wind took the crown fires and swept them through the timber, faster than a man could run. A crown fire generates its own wind, the hot air rising and the cool air coming in at the bottom of the trees, sweeping upward and fueling the resinous foliage with oxygen, and the fire turns into a fast-moving conflagration. We didn't talk about it, but I was planning our escape route; we could outrun a blow-up by moving downhill and across the slope. If the situation got too hot and violent, we could move into the burned area. It wasn't the best option, but we'd be better off next to the fire, where we could see what was it going to do.

We worked our way up the left flank just as the sun broke over the mountains. With the sun on our backs, the day got warmer fast. The trees that had burned in the night started to fall. We could hear the branches cracking and the crashes as the blackened trunks hit the ground in a shower of sparks and ashes. We kept our eyes open and were ready to move quickly if we had to.

"Let's go through the burn and get back to our packs and have breakfast," Carl said. He wiped his forehead with his sleeve, leaving a black smear. I was too tired to laugh. We must have looked like coal miners.

The ground leveled off into thick timber, still smoking, which only yesterday had been green, living trees. They were trees no more, just inverted black stakes. The grass and ground cover had been burned clean. Some of the logs were still in flames as we moved rapidly through the hot ashes.

"This fire went every which-way during the night," Carl said, high-stepping over the burn. "Hope our gear is safe." But the small clearing where we had left the jump gear and packs looked like any other black spot. All that remained were the blackened metal brackets of the harnesses and chutes. The jump suits were gone, the sleeping bags stringy ashes. Two piles of blackened canned goods, their labels burned off, were the only identifiable objects.

"Son of a bitch," Carl said. He rarely swore.

We were really isolated now. There weren't any other jumpers left in camp who knew about our situation. I wondered what we would have done if one of us had busted something on the jump last night.

"And so much for a secure place to camp," I said, digging around in the cans with the pulaski. "Let's see what we have for breakfast."

"The fire really ran last night," Carl said, knowing there wasn't a thing we could about it now. "This looks like a can of mandarin oranges," he said, tossing it to me. It was actually a can of Vienna sausages, which we both ate quickly.

"Have some canned ham. Can't mistake that." I rolled back the oval tin top. The bread tin was in the pile but the insides were charred. Carl found a can of warm peaches and we ate them for dessert.

We cleared a spot next to the rubble and sat down, then played a game of *guess what's in the can.* There wasn't a thing to do but lay back on the ground and laugh.

"We don't have a hell of a lot of anything. Our chutes, jackets and pants are ashes. The rest of the cans are half destroyed."

"And my shirt and toothbrush are gone."

"Toothbrush? Were you planning on going to a dance?"

"No, but my mouth feels like muskrat fur," I said.

"We may be here the rest of the summer." Carl poked around in the scattered pieces of equipment. "The only thing left of the other pulaski and shovel is the steel. The handles are burned off."

"Old Number 102 made its last jump," I said. "Just as well it burned up before it killed somebody."

"I did okay with 102, and I've jumped with it for years. It opened for me last night. It's the way a guy goes out the door. You have to have good body position or you can screw up any chute."

"It's gone now. Another devil parachute will take its place," I said. Working all night must have been making me feel this way — a little down, a bit punchy.

"Last summer a first-year man went out the door with the static line under his shoulder," Carl said, "and the opening shock damn near tore his arm off. That was his last jump. He didn't come back this year." The fear of being careless and making a bad jump never really left our minds.

I threw an empty black tin on the pile of rubble, and offered my own jump story. "Paxton made a training jump this year — his second — with the static line *under* the back board of his chute. He almost broke his neck. He said the hell with it and quit as soon as he hit the ground. We liked him — tried to talk him into staying, but his mind was made up. He said he was going into the Marines where it was safer."

"Some guys have it and some don't," Carl said, more talkative with food in his stomach. "You have to *want* to jump."

"I expected some help from McCall this morning, but I guess there's no one back in camp."

"You know what this means, don't you?" he asked, and I suspected he was going to dig me a little.

"What?"

"We'll have to take this fire on ourselves, one way or another," he said.

I came up on my elbows and looked past my dirty pants and boots. "Look at that smoke. It's moving. There's nothing in these woods but deadfall and bear grass."

"We'll have to try. We'll work one side toward the head, and hope some help arrives today or tomorrow."

"Tomorrow?" I asked. Carl was a man of action and it looked like he had a plan.

"Could be. They might send in a ground crew — fly them into Cold Meadows and ground-pound in from there. I'd expect them tonight."

"The bottom of the fire is contained, but it's not controlled," I said.

"Walking around the fire won't do much good. It's too big and would take too long." Carl stood up. "Let's get back to work."

We file-sharpened the pulaski and shovel. The time was seven-thirty. The sun was strong on the ridge and the heat of the day was beginning, so now was the time to make some progress, before the winds of midday.

We dug line through the bear grass all morning, and in the late morning we heard the distant drone of an aircraft. A seasoned jumper could identify a jump plane at twenty miles. This one had a smooth hum, so it wasn't the Ford Tri-Motor. Within minutes a silver single-engine plane came into view — the Noordyne Norseman. The Noordyne carried four jumpers. Maybe some help was arriving.

The plane descended toward the burned-over rim of the breaks. This looked like a cargo drop. We sprinted back along the edge of the fire line to the open ridge, as the plane came in at 200 feet. A small cargo bag, the type used to drop messages, small equipment or radios, was thrown from the open door. The miniature red parachute popped and the bag came down in the ashes. I ran to retrieve it, past the black twisted clumps of the whitebark pine where I had landed the day before. Inside the bag was a six-pack of Lucky Lager and a note from Wayne Webb, a squad leader and the parachute loft foreman. He had been in the Army Airborne in the big war. Wayne would later become second-in-command of the jumper unit at McCall and would retire from the Forest Service with over 300 jumps.

Carl and I read the note together:

ALL JUMPERS ON FIRES. WILL TRY TO GET MISSOULA JUMPERS IN TO HELP THIS PM. TWENTY-FIVE FORESTRY STUDENTS HIKING IN FROM COLD MEADOWS. SHOULD ARRIVE TOMORROW NOON.

WEBB

"How do we signal the Noordyne we need a radio?" I asked. A ground symbol was specified for that purpose, using the orange streamers, but ours were ashes. I put the note in my pocket and waved my helmet. Carl was cool as usual.

Wayne read our minds and saw the dilemma. By the look of the moving fire in the lodgepoles and the burned-over ridge top, he'd guessed the flames had overrun our camp. On the next pass he dropped a cargo bag with a radio in it. Most radios didn't survive the impact even with the chute deployed. This one didn't make it. We played with the controls but the tubes must have shattered.

The Noordyne continued to circle above us, and when they received no response, another message attached to a drift chute was thrown out.

WILL DROP FOOD AND SUPPLIES THIS PM FOR YOU, AND FOR GROUND CREW COMING IN.

HOLD THE LINE,
WAYNE

Carl picked up the message, read it, and waved back to the Noordyne. Wayne Webb waved from the door, and the plane turned to the west.

It was hot in the old burn, and the wind was coming up again. "Wayne said to hold the line," I said. "Let's do it."

"We'll need some help in here soon, either from McCall or Missoula," Carl said, picking up his shovel. "The fire's still rolling at the head. Let's just keep cutting through the bear grass."

"I could still use a drink," I said.

"The spring I saw from the plane *has* to be around here somewhere." Carl hunted the hillside until he found it, a little rivulet that bubbled out of the ground and splashed its way into the huckleberry brush, no more than a few feet away from an old elk-hunter's camp. Weathered wooden stakes were still in the ground where the tents had been set up, and cross poles were tied to trees for meat sacks. Rusting iron frying pans, buckets and a dutch-oven stove were piled under the tree branches. In a wooden cartridge box we found blue porcelain plates and cups.

"This looks like an outfitter's spike camp out of Chamberlain Basin," Carl said, rummaging in the gear. "Not much to burn here."

"Let's use the buckets to haul water and douse those burning logs," I said, thinking it would be a change of pace from swinging the pulaski.

We carried bucket after bucket of water to the edge of the fire and threw them on the flames. In the long run, the water wasn't as effective as shoveling dirt but it was fun listening to the hiss of the steam over the burning logs. After a day on the fire it didn't take much to entertain us.

We worked through the noon hour. The yellow and white smoke before us had doubled in size since the morning.

At one-thirty we stopped and walked back to the burned pile of canned goods to find something to eat. "We're just slowing this fire, not getting ahead of it," I said, twisting the opener around a can of plums. "We've only got a line around a third of this monster. We need more jumpers and a ground crew. Those forestry students will get some firsthand experience tomorrow."

Carl knocked the ashes from one of the last remaining cans. "They'll drop food and sleeping bags tonight for the ground crew. If there are extra sleeping bags we'll confiscate two of them. Either that or curl up in the cargo chutes."

"Just like a feather bed," I said, then got to thinking about home, and another world a long ways off. My mother would finish baking pies about now, and then maybe she'd sit down and write me a letter. She still thought I was working in a lookout tower. I'd tell her later I was jumping out of airplanes.

A snag cracked and fell inside the burn.

"A plane-load of jumpers is what we need to stop this hombre." I opened another black can: peaches. After a day in the heat, a guy craved fruit. The steaks could come later. "A two-hour nap would be nice right now," I said, looking out over the blue canyon of the Salmon River. Away from the smoke, the sky was crystal clear and the green mountains sparkled in the high sun.

"It would be," Carl said, leaning his back against a black log and dropping his hardhat over his eyes. "Five minutes. Then we go again."

The five minutes lasted fifteen, but we were revived and hit the fire hard. The pulaski and shovel continued to cut and swing their way through the lodgepole timber. We didn't have a crosscut saw, and there wasn't time to cut the deadfall, so we moved around logs, making snake-like lines in the dirt. At four o'clock we paused for a break, walking back to the small brook for a drink. And then we heard the plane.

"For chrissake, at last we're getting company," I said.

"If it was the Missoula jumpers they'd come from the east. Sounds more like the Ford Tri-Motor." Carl rarely used a four-letter word, but if there was ever a time to say something it was now. Carl looked west and said, "It's the Ford."

The three-engine plane came into view moments later and the pilot flew straight over the fire and began circling for a jump spot. The spotter selected the landing area, a small boggy meadow dotted with yellow flowers and ringed with small pines. The ground lay downhill and 400 yards off to the right flank of the fire.

"It's our guys. Hot blazes, I knew they'd get here."

"No reason to land on that burned-over ridge we used last night," Carl said, as if he were talking to the squad leader in the plane. "Let's go and meet them. There's the drift chute."

The pilot was on a serious mission. He and the jumpers knew this fire had to be stopped — it could blow up and roll across the lodgepole hills. The winds were over ten knots but not too strong to jump, and the plane was on a straight heading for the first pass of a two-man stick.

We had a quarter mile to cover to join up with the new squad. Two white parachutes blossomed in the blue sky; another two, and then two more until all eight jumpers had left the plane. From what we could see as we approached the meadow, they were all landing easily in the soft grass. We closed on the meadow, swishing pine branches back and sidestepping the huckleberry brush. We wanted to see the new guys, but we weren't going to admit to them we needed help on this fire, which had increased to more than several hundred acres.

At the edge of the grass, a smokejumper stood next to his parachute bag and firepack. He was slim-waisted and wore faded blue denims and a short-sleeve khaki shirt. He was six feet tall, thirty-two years old, had blonde hair and teasing blue eyes. His features reminded me of Richard Widmark in the movie *Red Skies of Montana.* His forearms were tanned, thick, about the size of small tree trunks. A Bull Durham tag hung out of his shirt pocket. He had the look of authority about him as he stood there, one foot on the stuffed parachute bag. His name was Richard A. Peterson, but he was known to every smokejumper in Idaho and Montana as Paperlegs, a name he got from somewhere but no one dared ask. Paperlegs had been stepping out of airplanes since 1947 and would continue to do so for the next twenty-five years.

He looked well rested and ready to go as we walked up, ash-covered and dirty, our eyes bloodshot and hollow after twenty-four hours without sleep.

"I thought you guys had a real fire going here," he said, looking around the meadow. "My grandmother could stomp this out with her boots." Just then, three tall pines behind him ignited in a pointed column of black and yellow smoke. Paperlegs had a way of talking to you, a brusque exchange, followed by a short laugh. Every smokejumper knew and respected him.

"If we hadn't been here this fire would be burning into the Middlefork country by now and wouldn't stop until the first snow flies," Carl said, swinging his hand in the direction of the burned rim of the canyon.

"And I heard you lost your sleeping bag and chutes."

"We were working the bottom in the canyon, and the fire ran across the top of the ridge last night," Carl said as he pushed his shovel into the loose dirt, looking more relaxed.

"We survived," I said. "There's some black cans out there we're still opening."

Paperlegs pulled the pack of Bull Durham out of his pocket, peeled off a paper with two fingers and leveled tobacco on it with the other hand. He licked the paper with a flourish, rolled it, and with a quick stroke of a match over the back of his pants, all in one motion, lit the short cigarette.

I made a mental note to practice that move.

The jumpers in the meadow assembled their gear, and one of them carried it to the edge of the trees.

Paperlegs took a long drag on the Bull Durham. "Reid didn't have to call Missoula. A planeload of us got back from the Hat Creek fire up by Challis. I jumped with Ellis on the first stick, Spence Miller and Ray Beasley jumped the second stick, and Donnelley and Marker are coming up now. Lonnie Park and Gene Lewton jumped last. We'll have this thing out in no time." Paperlegs laughed again when a tall pine burst into flames, like someone had thrown a jug of gasoline on it.

Miller and Beasley walked up. Their black suspendered logging pants were clean and Beasley's even looked creased. Ray Beasley was the best dresser in the outfit. Their blue work shirts showed only the dust of the jump. Our clothes were blackened and gray from ashes and dirt, stiff with dried sweat, and spotted with small holes from hot embers.

Ray Beasley nodded a greeting and pulled a meerschaum pipe from his pocket, tamped it and lit up. I wondered where he kept the pipe when he jumped and why it wasn't shattered into a dozen pieces.

Spence Miller, a dark handsome man, stopped and with a precise swing of his arm, sunk the axe end of his pulaski into a weathered log.

"I thought this fire was under control," Spence said, trying to get our dander up.

"It is," Carl answered, smiling at me. "You guys are just here to mop up."

"Mop up, my ass," Spence said. A line of trees on top of the low ridge was aflame from ground to top for a hundred yards along the crest. "This fire is on its way to Montana," he said.

Just then Gene Marker and Bob Donnelley walked up, pulaskis and shovels in their hands. Marker, athletic and suntanned, with a perpetual mischievous grin on his face, started singing, *"I ride an old paint, I lead an old band, and I'm going to Montanee to throw the hooleehan."*

"Just hum a few bars," Beasley said.

"Don't encourage him," Bob Donnelley said. "I'm going to throw up."

Gene Marker, still grinning, sat on a log watching us, his hand holding the lady shovel. "You guys already got the two-thousand-yard stare. Lookit, Bob, don't they?"

I had to agree — we probably had the stare, a gaunt, hollow-eyed look common to firefighters too long on the line.

"By gosh, they do," Bob said. "We thought you guys were sleeping out here."

Lonnie Park and Gene Lewton came through the trees and greeted us. Seven out of eight jumpers were accounted for. Gene Ellis was still missing.

"If you jumped with Ellis, where is he?" Carl asked Paperlegs.

"When I left him, he was looking for his cigarettes. He said they flew out of his pocket on the opening shock. He's probably on his hands and knees somewhere back in the trees."

Gene Ellis walked up to the group, looking at the men with his black piercing eyes. He was a college boxer and a tough jumper. He wore stagged pants with red suspenders, and his metal hardhat was tilted at a jaunty angle. The hat had been punctured by two small-caliber bullets. Somebody had used either the hat or Ellis as a target.

Ellis's first words were, "You got any hot coffee?"

"We don't even have cold coffee. The fire took out our gear and supplies last night," Carl said. He had been working too long on the fire to get worked up over Gene's request.

"I figgered you'd muck up the cooking. I brought some coffee in my pack. Help yourself."

"It's too hot for coffee, anyway," Carl said. "Let's get this fire out."

All the stories about Ellis were true. If you were ever on a fire with him and he forgot his coffee and cigarettes, you'd probably have to kill him.

Paperlegs broke up the conversation. "We have work to do. Carl, you're the Fire Boss. You know the rules — first senior squad leader on the fire calls the shots."

"All right, I'll take it. We have this smoke about out anyway."

"Sure you do," Paperlegs laughed.

Carl took over quickly and quietly. He divided the eight new jumpers into two teams. Paperlegs was to take one team across the burn and work the south flank. Ellis would take the other team and cut line on the north, both moving to pinch off the fire at the head. Carl and I would double back along our line and retrench the bottom of the burn in the canyon. The ground crew, when they arrived, would continue widening the line, cut down smoking snags, and mop up the hot debris. When the fire was under control, the jumpers would leave and be the first on other fires.

Paperlegs was directing his team. "If this fire blows up, we'll have a lot of work to do, so let's get it out now," he said. "The ground crew will be here tomorrow. We may have to work through the night."

Not another night without sleep, I said to myself. I was starting to catch the heels of my boots stepping over logs, not a good sign. And the pulaski kept getting heavier. Carl looked like he could go another night if he had to.

Carl solved the sleep problem. "We'll get a food drop tonight. Let's work the hot spots and the lines till midnight. That's the plan for now. When you hear the plane, gather back here."

Thanks Carl, I said to myself. I think I can stay awake at least until midnight.

The Ford Tri-Motor came in just over treetop at six o'clock, all three engines at full throttle, the pilot not taking a chance on stalling. The plane made eight passes and on each pass the cargo handler kicked out a large gray insulated container, delivered to the ground by a white nylon chute. There was a hot meal of roast beef, boiled potatoes, gravy, corn on the cob, packs of iced tea, coffee and ice cream, and sleeping bags for the ground crew. The ground pounders carried their own shovels and pulaskis.

The last drop contained a radio, and this one worked. Carl communicated with Del Catlin, who'd been kicking the cargo from the plane.

"We should have the fire controlled tomorrow. The ground

crew can mop up."

"If you can do that, send Paperlegs and the other seven guys out Wednesday," Del said. "We need them back at base. You direct the forestry students. Don Miller, the fire guard at Chamberlain, is bringing them in."

Del continued, "Dub Horn will bring in a string of mules to pack out the gear when you give us the word."

"Okay. By the way, what's the name of this thing?"

"They're calling it the Lynx Creek fire."

Carl signed off. We quickly ate a hot supper and went back to the line. I felt much better after a hot meal, but hitting the same place twice with the pulaski began to take more concentration. If this fire doesn't blow up, I thought, we could be in bed at midnight. The hell with even looking for a sleeping bag. At that point I just wanted to lie down.

The moon had come up and by the time Carl and I had made the swing around the bottom, it was one in the morning. Almost two days without sleep. I did find a sleeping bag at camp, fell into it without even looking for the zipper, and was asleep in an instant. I never noticed the rocks underneath until the next morning.

Fighting the fire at night, when the air was cool, brought it under control. The forestry students arrived on the third day and attacked the burning snags and widened the line. In the end, the Lynx Creek fire had burned 300 acres of timber. It could have been catastrophic if the fire had blown up in the old deadfall and rolled over the hills of lodgepole.

At noon on the third day, Paperlegs and his squad assembled around the cookfire and announced they were moving out. "Reid wants us back at McCall. We'll hike to Cold Meadows and be there by nightfall," he said. "See you back at the base."

"We're not in love with this burned over ridgetop," Carl said. "Ask them to send in food for three more days."

"You're the fire boss," Paperlegs said, laughing. "Maybe with luck, you and Bud will get out before the end of summer."

"We'll ask Del to drop a couple of six-packs on the next cargo flight," Beasley chipped in, lighting his meerschaum. "That should last you until September."

"You'd think they wanted us to homestead this place," I said.

Gene Ellis scratched a match on the back of an iron frying pan and lit a Lucky Strike. He reached over the fire to the grill, confiscated from the elk hunters' camp, and poured himself a cup of coffee. "One cup for the road. Don't break a leg." Ellis downed it, nodded, and followed the jumpers down the trail into the forest.

The strategy of fire control had worked. Our three crews hot-spotted and cut line around the flanks and linked up at the head of the fire in the lodgepole in the darkness. The forestry guys widened the line and covered the smoldering embers with dirt. They felt the burned-out stumps with bare hands to be certain any remaining embers would not ignite when fanned by new breezes. On Friday, they and the Forest Service guard hiked back out to Cold Meadows.

It was the seventh day on the burn, and I pulled on my cold leather boots. The sun broke over the mountains, lighting up the fog on the river below us. It looked like a white lake. "Why are we still here?" I asked, to no one in particular.

"Because we were here first," Carl said, sliding the pans around for the morning campfire. "We'll make sure every smoke is out."

Any real jumper wanted to work as many fires as he could in one season. It was a toss-up between spending lots of time on large fires and adding up the overtime, or making many jumps on small fires. In the long run, making more jumps seemed to be the most important. Why? To increase our jump tally, to go to more fires, and to keep the excitement going. Maybe that was it. The squad that had left on Wednesday had probably already jumped again. I could see Ellis digging line, drinking coffee and smoking Luckies on yet another burn.

On fires of this size, a government horse packer came in and took out the equipment. Dub Horn, the packer, was arriving that day to pick up the parachutes and equipment. Sometimes he'd arrive on the fire a day or two after we did, and oftentimes a week later, depending how deep in the wilderness the fire was burning.

At noon as we sat on cut stumps around the campfire, a tall man riding a buckskin horse came quietly up through the pines. The buckskin stopped ten feet from the outside circle of the camp. A string of ten mules behind the lead horse stopped, as if on command, drooped their heads and begin flicking their ears at the summer flies which had caught up to them.

Walter C. Horn, better known as Dub, was six feet, six inches tall, his faded Stetson a shade lighter than the big horse he was riding. He was so tall that he just swung a leg over the pommel and stood on the ground. He sauntered over to the campfire.

"You boys plan on wintering up here?" Dub asked. "You been up here quite a spell."

Indeed, it was Dub Horn himself, a legend in his own time. It was said he was the best packer in the Rocky Mountains, and he was certainly the tallest. He controlled his mules, the legends said, by sheer fright: he was bigger than they were. He could pack a mule — I'd seen him do it — by tying down the manty on one side, throw the diamond hitch, then reach over the mule and tie down the other manty, all without moving his feet. If a mule got ornery Dub would hit him between the eyes with his fist. I'd never seen him do *that.* But I didn't doubt those stories for an instant.

"We're not planning on it, but another week up here and you can tell the bosses at McCall that we need a mule and some seed potaters." Carl said shifted his position on the stump to get out of the campfire smoke. He looked dirty enough to be a potato farmer.

"I'll radio that message when I get back to Cold Meadows," he said. After Dub finished the sandwich and coffee we gave him, he loaded the packs, the mules all the time giving him the wary eye, but nonetheless holding steady when he'd throw the packs. Once in a while one would lay back his ears, but Dub cuffed him on the rump. Twenty white packs were neatly secured on ten mules. Dub swung into the saddle and led the string away, every one of them turning and kicking up dust just outside the ring of the campfire. The lead mule hee-hawed, answered by a big black. Dub's horse whinied, as if to tell them to shut up, and Dub waved and moved down the trail. We watched the mules until we could no longer see them in the trees. Dub must have passed on our request because the

Noordyne came in the next evening.

The pilot was well acquainted with the camp on the open ridge and the small cargo bag landed right next to the cookfire. In the bag were four T-bone steaks, two cans of baked beans, and a dozen large baking potatoes. There was a note in the bag.

TO THE HOMESTEADERS-
WE CAN'T SPARE ANY MULES. HERE'S SOME POTATERS. COME OUT TOMORROW. A PLANE WILL PICK YOU UP AT 6 PM AT COLD MEADOWS.
WEBB

After nine days, we took one more look at the burned timber, picked up two cans of peaches for the trail, and hiked to Cold Meadows. Exactly at six o'clock, Bob Fogg picked us up in the Travelaire. That same evening, after showering and changing into clean clothes, we hoisted beers over the bar at the Foresters Club. The Lynx Creek fire was another adventure behind us. We had held the line.

The cute redhead was playing the slot machines. Two guys at the other end of the bar watched her. They didn't look so tough, I thought, analyzing the situation just in case she gave me another smile. But mostly our conversation over the wooden bar was about when the buzzer at the parachute loft would sound off, sending us to the fire cache to pull on our jump suits and climb back into a plane and take off to another smoke.

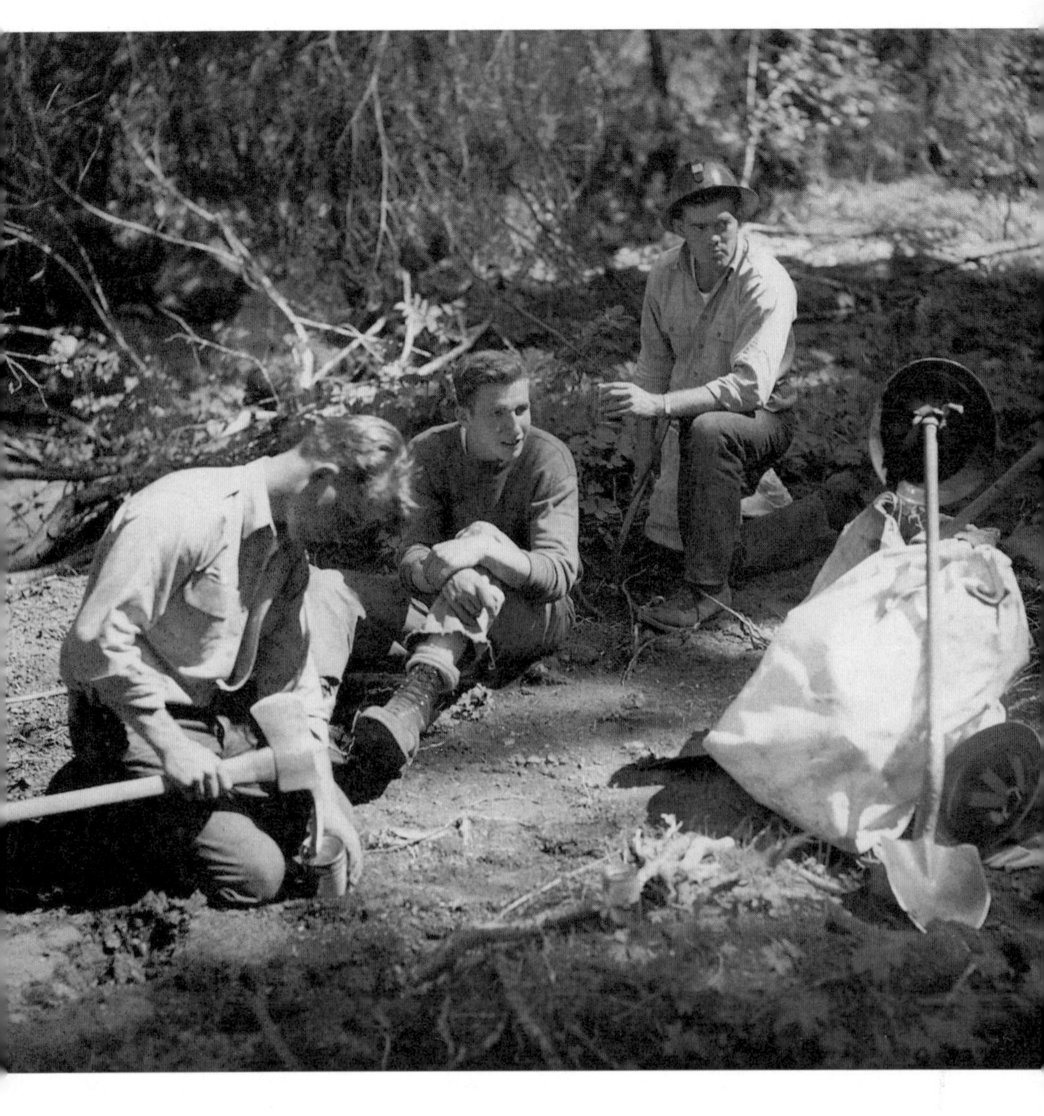

The fine art of operating the pulaski can opener

Chapter Eight

OFF TO BIG CREEK

We were settled in the bunkhouse, listening to each other's jump stories, when Reid Jackson came in and announced assignments of project work on ranger districts. "Carl Rosselli, Bob Rakozy, Larry Clark and Bud Filler," Reid said. "You guys saddle up for Monday-morning departure to Big Creek." Reid rattled off several other projects but the magic of the word "Big Creek" ran through my head and I heard nothing else after that.

Big Creek ranger station was east of McCall, in the Idaho primitive area. It was accessible by flying in, or by driving 150 miles over a winding unpaved mountain road in a four-wheel-drive vehicle. There was a second route, longer but more favorable, following the river canyon along the South Fork, up the East Fork of the South Fork, to Yellow Pine, and then over Profile Summit to the ranger station. Both roads were closed after the first big snowfall in October or November, sealing off Big Creek. After that any inhabitants of the valley had to depend on snowshoes to get out.

The ghost town of Warren lay north of McCall, just about halfway to Big Creek. In the fifties the population of the town was just a few folks, either hermits who liked the solitude or a couple of bartenders selling drinks and telling tales in the Warren saloon. In 1862, when gold was discovered in the Warren Valley, the population jumped to 2,000, all working to make their fortunes in placer mining. Most of the labor was carried out by Chinese workers, many of them at rest in the cemetery behind the Warren saloon. Fortunes were made by the early miners, and when their claims played out, the diggings were purchased by the Chinese, who made a second fortune, working for nothing, laboring in the tailings.

When these claims were abandoned, a few Chinese men traveled east across two 9,000-foot mountain ranges to the valley of Big Creek, and even beyond. But mining was difficult there and hauling the ore out of the backcountry was a long and dangerous task. The winters were brutal. If laboring in Warren was tough for the Chineses, mining in the Big Creek valley was back-breaking. There were many stories told around the campfires of their hardships. Reid Jackson, who frequented the valley in his early smokejumping days, wrote a poem about their misfortunes. Slim Vassar, the fire dispatcher in McCall, and Val Simpson, the ranger, after a few shots of snake bite medicine, swore the story was true.

"Come on, Reid, give us the poem," we said. He smiled, and we knew he was getting ready, so we cracked a couple of beers. He had it memorized.

A PACK-STRING STORY

This is a tale that was told to me
by a ranger of long ago,
of the mining days and men's strange ways
in the wilds of Idaho.

'Twas back in the days when roads were few,
and travel was mostly by trail,
so, gather 'round and listen close,
and I'll tell you that ranger's tale.

'Twas early in the springtime,
when the snows were deep and cold.
The trails were mostly buried,
as the miners sought for gold.

Word came out from a wilderness camp
to authorities in the town,
an avalanche had killed ten men.
Ten Chinese had been put down.

A contract had been drafted
by the mining company boss,
with a note to the county sheriff,
informing him of the loss.

The contract was for two packers,
and a pack string plus the tack,
to come and get the miners,
and bring their bodies back.

The sheriff would send the bodies
to Seattle on a train,
where each would be cremated,
and to China would go the remains.

Now, Donahue and Murphy
were two cowboys out of work.
They owned a string of ten good mules,
and a challenge they'd never shirk.

Their bid was fast accepted, and
they saddled up their mules.
They mantied up some groceries, and they
gathered up their tools.

They mounted up their horses,
and they headed up the trail
to the mining camp so far away,
and the north wind it did wail.

The trail was difficult at best.
The snow was awful deep.
They often had to stop and rest,
and the cowboys longed for sleep.

They pushed through mountain passes,
and canyons deep and blue.
They worked around the blowdowns,
and they sometimes threw a shoe.

They crossed many icy rivers,
and they watched for snow slides too.
And the north wind made them shiver,
as they stopped to eat some stew.

After many miles of travel,
they reached the mining camp.
The boss was there to meet them.
It was dark, but he had some lamps.

He showed them to a cold warehouse,
where the Chinese men all in a row,
were all laid out in grotesque shapes,
just as shoveled out of the snow.

They were frozen stiff and wouldn't bend,
they presented an awful scene.
And Murphy said to Donahue,
"This would make a packer scream."

He said, "We need to find a way,
to lash each body on a mule.
The trouble is they're frozen hard,
and can't be bent in two.

I suggest we take an axe
and cut each body in the middle.
I suspicion that's the only way
that we can solve this riddle.

With half a body on a side
each load should balance well.
In fact I rather think that we
could pack them clear to hell.

It really doesn't matter
that they won't look neat and pretty,
they're only going to burn them up,
when they get them to the city."

Now, Donahue was squeamish.
He didn't like the plan at all,
but he went and brought the mule string in,
and tied them to the wall.

Murphy did the axe work.
It was an awful task.
He never could have finished it
without his whiskey flask.

With the axe work all completed,
and after a good night's rest,
Donahue saddled the mules again,
they were ready for the test.

The air was strangely heavy,
though it was forty below,
both packers were drenched in sweat,
and were moving rather slow.

The mules were awful skittish,
as nervous as could be.
Each was watching Murphy,
to see what he could see.

Murphy tightened up the cinch,
of the first mule in the string.
He picked up half a Chinese man,
and tied him with a sling.

But his stomach got the best of him,
his complexion turned quite pale.
His gorge was working overtime,
his composure was about to fail.

As Murphy turned to vomit,
he swears, so we know it's true.
that all ten mules in the pack string,
were busy throwing up too!

Now, this is the tale that was told to me
by that ranger of long ago,
of the mining days and men's strange ways,
in the wilds of Idaho.

I pictured Big Creek, the mountain passes and the whitewater rivers, tumbling through the forests. I could hardly wait to get there. If the place was as isolated as they said it was, I'd love it.

Our four-man crew had the job of opening the road past the town of Warren, down across the South Fork of the Salmon, up Elk Creek, over Elk Summit, and down into the valley of Big Creek itself. The trip in would take all day, clearing trees fallen over the road from the winter storms. We'd dig out rock slides and shovel away the remaining snow banks on Elk Summit. It was good duty, and when there were no fires, it beat piling brush.

"Going to Big Creek is the best of the project work," Bus Bertram said, leaning back on his bunk and directing his comments to me. He had a leer on his face. Bus was ten years older, a veteran from the big war, a jumper from 1947 who stayed off several years, and then came back. He was as smart and as tough as they come. The rumor was he never lost a barroom fight. If he didn't like you, you didn't fool with him. When he and Paperlegs were together people gave them a wide berth. I knew something was coming.

"Take your sport coat, a white shirt and your rep tie," Paperlegs said. "There's a girls' college right near the ranger station."

"Yeah, and the women are lonely," Bertram added.

"Plan on being there the rest of the summer, " Paper grinned. "We'll be on fires soon and the bosses will forget you guys in the backcountry."

There was envy in their ribbing. Anything was better than piling brush and painting buildings. Carl Rosselli and Larry Clark were happy with the assignment. Larry had the ruddy face and crewcut of a football player, which is what he was, and he joined in: "They want the brush cleaned up from Warren to Big Creek. They said we were the best squad for the job," Larry said. He admitted he always got the brush-piling assignments.

Bus leaned back in his bunk and lit a Lucky Strike. "Now don't forget your sport coat. The dances at the girls' college are formal."

"That's also the country of the sidehill gougers," Larry said.

"Sidehill gougers," I said. It was a comment and not a question, but I knew I was going to learn about them anyway.

"Yes, sidehill gougers," he continued. "I've been in there and I've seen them. Big furry animals, mean and vicious. They live high up on the rocks."

Bus and Paperlegs waited to see if anyone rose to the bait. "The mountains are so steep at Big Creek that the gougers' legs on their downhill side are a half-foot longer than their uphill legs," Larry said.

Carl Rosselli, who had been there before, shook his head and started to pack his duffel.

I turned to my locker and threw some work clothes into a pack. *Big Creek—the land of the frozen Chinese miners, sidehill gougers and lonely women—this was going to be an exciting trip.*

The next morning the four of us left in a four-wheel-drive Forest Service pickup, two in the front seat and Rakozy and me in the back, straddled among the boxes of food, shovels, axes, chains, cross-cut saws, come-a-longs and personal duffels. Ten hours later, after crossing the South Fork, we had shoveled and chopped the road out to Elk Creek Summit. This was wild, steep country, with dark stands of virgin fir and spruces in the draws, and open granite clearings on the ridges. Land that never would be developed or logged, I thought, but country that *could* be destroyed in a crown fire. Three to four hundred years after such a fire, it might, just might, look like this again.

In the late afternoon we dropped into the Big Creek drainage. Lodgepole pine grew thick in the valley, especially next to the creek, the home of summer-spawning salmon. From the switchbacks of the road, we could see the creek itself through breaks in the timber, splashing its way north to the distant canyon of the Middle Fork. Most of the mountain peaks were above timberline, with patches of snow, even in July, resting on the northern slopes. Goat Mountain loomed like a citadel at the head of this high, beautiful valley.

Several Forest Service buildings had been erected at the edge of the timber: the station itself and a small log cabin where ranger Ted Koskella and his wife lived. The assistant, or alternate ranger, Bob Dustman, and two summer employees lived in quarters next to the corral, just abeam of the uphill end of the grass airfield. The runway itself paralleled Big Creek. The airstrip had a dip in the center and any pilot landing there had to adjust the touchdown point to land before the dip. The only planes using the airfield then were the high-wing tail-wheel aircraft with powerful engines. The Travelaire was well-suited for this work. It was used to ferry firefighters in and out of Big Creek, but only in the morning and evening hours. Landing at mid-day at the high elevation and in gusty winds was courting disaster.

A gray bunkhouse was a half-mile north of the end of the runway. Built by the Civilian Conservation Corps in the late thirties, it was used by the crews who opened up this country, and was where we bunked and cooked for the length of the project work.

Ted Koskella had an enviable job, at least it seemed to me, far from the administrative work of the Payette National Forest at McCall. Ted, his family, and the other government personnel, as well as the caretakers of the lodge, left the valley just before the snows closed off Profile Summit and the settlement of Yellow Pine. Elk Summit would have long since been snowbound and inacessible.

A quarter-mile south of the ranger station, along the narrow dusty road, was the privately owned Big Creek Lodge. The two-story structure was built several decades before of native logs, now weathered gray, and many well-spaced white trimmed windows. Like the Forest Service buildings, it was kept in good repair. A covered porch on the front looked over a pasture across the road. Beyond the pasture, a line of pines marked Big Creek, and beyond the creek a steep hogback ridge rose 500 feet from the valley floor.

Granite slopes of rocks dropped from the hogback to the creek bottom, and a faint game trail snaked its way across the rocks just above the creek. In the evening, from the porch of the lodge, we would often see a deer or elk walking along the trail from one thicket of lodgepole to the next.

The proprietors of the lodge catered to the occasional traveler from Warren or Yellow Pine, and to hunters in the fall. An outfitter based at Big Creek packed hunters down the creek in the direction of the Middle Fork, or across the saddle to spike camps in Chamberlain Basin.

The businesss changed hands frequently since the few summer visitors and the hunters didn't provide a very stable income. Those who owned the lodge loved the backcountry and the solitude of the valley. The managers, Mr. and Mrs. Holden, lived in the back. The main room served a multiplicity of purposes. The fireplace was in the center, a heavy plank dining table on one side next to a comfortable leather sofa, and on the opposite wall, a glass display case where a traveler could buy a box of .22 shells, some trout lures or a candy bar. The food served in the lodge was excellent, but simple fare: bacon, eggs, potatoes and hotcakes in the morning; hamburgers, salads and lemonade at noon; and thick steaks, mashed potatoes, and corn on the cob in the evening—maybe even a blueberry pie or two.

It was a mile from our bunkhouse, past the ranger station, to the lodge, along the single road that paralleled the airstrip. The road went straight under the canopies of the pines until opening up in the meadow.

There was, of course, no women's college there, or even the sight or scent of a woman. Visitors were rare. The Forest Service personnel, the managers of the lodge, and now the four of us made up the settlement of Big Creek. If we had to work on projects for four weeks before the fire season got rolling, it was good duty, especially since I loved the wild, mountainous country.

"If you guys run out of things to do in the evening," Ted Koskella said, "stop in at the lodge. Mr. and Mrs. Holden are running the place and Mrs. Holden bakes some great pies." The other jumpers didn't pay much attention to Ted's suggestion, but I did—I was a hound for homemade pies. Anyway, I might get tired of

Rosselli's cooking and a piece of pie could save the day. I'd check out the lodge later.

In the morning, when the air was cool and the sun had not yet touched the high peaks, small patches of fog drifted ghostly across the meadow, which was also the runway. Almost every morning, three and sometimes four black bears came out of the timber and played in the meadow. Eventually, as daylight came on, the bears disappeared into the trees by the creek and fished for salmon.

By the third morning, we were used to seeing the black bears from the windows of the old bunkhouse. Carl Rosselli, standing by the window, called out: "Look at the new bear out there! The big one with the hump, over near the woods, apart from the others."

"That looks like a grizzly, boys," Bob Rakozy said. He unsnapped the tiedown on his holstered .45.

"It sure as hell does," I said, watching the bear swaggering toward the creek.

"They're not as big and mean as sidehill gougers," Larry Clark said, picking up the keys for the truck.

We watched the bears for another ten minutes until it was time to go to work. The buff-colored bear, larger than the others, kept his distance from the blacks as they all made their way to the creek.

Ted Koskella greeted us at the ranger station. "That might have been a grizzly in the meadow. I saw him from here. The big bears come in for the salmon. Pretty rare here, though."

"We'll stay away from the grizzly," Carl said. "The black bears won't bother us, but the grizzly might."

"He'll be hanging out in the woods during the day," Ted said. He was grinning. "Probably somewhere in the trees next to your bunkhouse."

Carl and Larry went off with the ranger to repair telephone lines, and Bob and I spent the day digging a trench for a water line from the station to the road. The hard exercise in the hot sun felt good, but the work was repetitious and my mind wandered. Tonight was the night to check out the Big Creek Lodge.

After a supper of ham, boiled potatoes and stewed tomatoes, à la Rosselli, I washed up and walked the dusty mile to the lodge.

Evening was fading. A hawk circled in the thermals above the hogback. All seemed quiet at the lodge—nobody nearby or on the front porch. The only movements were the horses across the road in the pasture, swishing their tails at the flies. I didn't think it would be intruding if I stepped inside. It was, after all, a place where meals were served. I walked across the plank deck and stepped inside, shutting the door behind me.

The room was filled with the warm smells of wood smoke and baking bread. An elk rack hung over the stone fireplace, and a big set of mule deer horns reared up from the pine paneling on the opposite wall. In the back, a dining room table was set with heavy china on a white tablecloth. A leather sofa and a few stuffed chairs surrounded the fireplace, undoubtedly the center for drinks and hunting stories over the years. Off to one side, a glass display case housed the candy bars and fishing paraphernalia. A single door opened to the kitchen, and the sweet smell of fresh pastry seeped from around the door.

Maybe Mr. and Mrs. Holden were in the back, I thought. Somebody in there was doing some great baking, and I was getting hungrier by the minute. Maybe I should leave now, I thought. Well, this *was* a store and I wasn't intruding, and a piece of pie would taste good.

I stood in the center of the room, trying to decide what to do, when the kitchen door opened. I expected to see a short rosy-cheeked cook with a white apron and flour-covered hands, but standing in the doorway was the most beautiful young woman I had ever seen. Her eyes were large and brown, her hair cut short and the color of a fawn. She wore a white cotton blouse and faded denim shorts, and her legs were long and slim and tan. She was tall, but I figured I was just a little taller. The way she stood there and smiled, her head with a certain aristocratic tilt to it, stopped me in my tracks.

"Hello," she said, and her voice was smooth and her eyes were perfect, and she was, even in the backcountry, wearing a trace of red lipstick.

"Yes, well, the door was—well, hello," I stammered.

"Can I help you?" she asked. She walked to the edge of the counter, and it was as if I had never seen another woman before.

The rest of the room faded from sight, and I felt like I could float over to her.

"Yes, I was wondering if you were open."

"We are," she said. "You surprised me. We don't get a lot of visitors this time of year. There'll probably be some fishermen in soon from Warren. A crew opened the Elk Summit road two days ago."

"That was our crew. We opened the road on Monday," I said, and smiled back. I wanted to say the right thing to her, and not screw up. *God, she was beautiful.* "The ranger said you make some great pies, and I thought I'd find out," I said, still smelling the baking from the kitchen.

"Of course," she said. "My mom just made some blueberry. I'll bring you a piece."

When she went into the kitchen, I watched her movements until she went through the open doorway, then turned and looked at the fishing lures in the glass case. I loved to fish, but I wasn't thinking of lures then. There they were, those red and white and yellow trinkets with treble hooks attached to one end, displayed on pieces of heavy white paper. I glanced back to the kitchen door.

She returned and placed a quarter of a blueberry pie on the table, along with a glass of milk. "My name is Liz," she said, looking directly at me.

I introduced myself. For a moment, the pie was forgotten. In fact, at that moment, just about everything, except Liz, was out of my mind.

"My mom and dad manage the lodge," she said, and as if in answer to an unasked question, continued, "I'm helping them for a month before I go back to Boise."

"That's great!" I said, thinking of the three weeks we'd be there on project work, then paused, not wanting to sound overly enthusiastic. I started to work on the pie.

"I came in last Sunday," she said and smiled again. "My brother's a pilot and he flew me in." For a moment she looked out the window, then back to me. "I love it here. I love the mountains. I love to fly. I wish I could stay here forever."

"Our crew will be here for three weeks, unless we're called out to a fire sooner," I said. "This pie is great." I wasn't making much headway on finishing it. I wasn't sure of what to say now. She was very pretty, eighteen or nineteen, I thought.

We sat there and talked, maybe both of us trying to find the correct words. She asked me out to the porch, where we sat on the wooden steps.

We were trying to find out as much about each other as we could, and hoping the questions and answers were the right ones. We discussed the valley, she talked about flying in and the beautiful rugged mountains her brother showed her, and I told her of driving in, and then both of us wished we could stay for the whole summer. An hour had passed, but it seemed like minutes.

"I better be getting back to the bunkhouse," I said. "Can I come back and see you tomorrow evening?"

"I hope you will," she said. "I'd like that." Her eyes held more than a hint of excitement and mystery. "I'll show you around Big Creek."

"That will be fun—to see Big Creek." I took two silver dollars from my jeans pocket, opened the door, and laid them on the counter.

"The pie is on me," she said, and picked up the two dollars and placed them in my hand. Her touch seemed to linger a moment, and my heart jumped. I wanted to hold onto her hand, but I didn't. I turned for the door, and couldn't help but notice how smoothly she walked across the room. Her legs were just perfect, and she had that little tilt to her chin, like she was sure of herself.

"It's a mile walk back," I said. "I'll see you tomorrow evening."

"Tomorrow evening," she said, and gave me a short wave.

I turned down the road in the direction of the old CCC camp. The night was almost pitch black, but a trace of the moon was beginning to show over the high peaks.

We hadn't really gotten very personal, but every word she said was going through my mind. I walked past the ranger station and was only thinking of Liz as the road entered the black timber. The moon had cleared the hogback ridge. I was walking, almost skipping. Was it the dark road in the timber, or the thought of seeing her again?

Then I remembered the grizzly bear we'd seen that morning from the bunkhouse window, and the ranger said it was hanging out in these woods. Hell, he was just pulling our legs, throwing us a scare. Along the dark lane, the trees cast wavering shadows, long and black, and before long I was certain a bear lurked in every shadow in every dark hollow. A jay squawked back in the trees—jays usually didn't make noises at night.

If a bear were out there, it wouldn't bother me. Just then I felt I could take one on bare-handed. The jay squawked again, a sound of alarm, and just to be sure, I picked up the pace. When the lights of the bunkhouse came into view through the trees, I breathed a sigh of relief.

I walked in and slammed the door with a kick of my toe. The bunkhouse smelled of wood smoke and lantern fuel. Rosselli and Rakozy were asleep. Larry Clark was waiting up for me, I suppose, reading another western novel. When he heard the door, he slipped his glasses lower on his nose and glanced up. "Where in the hell have you been, Fearless? Didn't run across that grizzly, did you?" He took off his glasses and laid the book on the floor.

I gave a short leap and landed on my back on the springy bunk. "No," I answered, "I didn't see the grizzly, but I did find the girls' college." I laughed, because I had a secret, then rolled to one side and turned down the mantel on the hissing lantern. Larry gave me an inquisitive glance as the Coleman fluttered and went out.

The moonlight poured through the window of the old bunkhouse. Outside, the night sounds started. A breeze stirred the branches above the roof and the smell of pines came in on the night air. I thought of Liz and of tomorrow, and midnight passed before I finally fell asleep.

Bill Yensen

Chapter Nine

LIZ HOLDEN

Two riders came up the road, one on a bay horse and the other on a black mare. The taller one was Jack Kooch, supervisor of the Payette National Forest, and his partner was Ted Koskella, the Big Creek ranger. They stopped in front of Bob and me, dusty and sweating and stripped to the waist in the hot sun, still working on the drainage ditch from the ranger station.

"Howdy, boys," Mr. Kooch said, swinging down from the bay. Tall, lean and rugged, a tough ranger from the old days, Mr. Kooch was only a few years from retirement. He was respected on the forest and his word was law.

We took a break from the shoveling. "How was the fishing?" I asked.

"Oh, not too bad," Jack Kooch said.

Ted untied a brown rucksack from his horse, and with an exaggerated grunt, lowered it to the ground. He opened the pack, pulled off a light canvas cover and displayed ten of the fattest rainbow and cutthroat trout I had ever seen. Some were over two feet long and as wide as the salmon we had seen in the shallows of Big Creek. The sun glistened off their still-wet sides. Ted smiled and wrapped them back up in the canvas.

"I have the feeling you won't tell us where you caught them," Bob said.

"Nope," Ted said. He was smiling but I knew he wouldn't tell.

"Up there in those lakes," Mr. Kooch offered with a broad sweep of his hand, taking in an arc of 360 degrees, in as vague a fashion as any self-respecting fisherman could muster.

The mountains surrounding the Big Creek Valley were filled with high lakes, all habitat of trout of varying sizes and species—goldens, brooks, rainbows and cutthroats—and most of those lakes were undisturbed from summer to summer. They were difficult to find, and many of them took more than a day to reach, either afoot or on horseback.

The lakes were near timberline, hidden at the base of the granite slopes coming off the peaks. I didn't know their names but most of them would show up on a Forest Service map. Now I had a challenge: to find the secret lake, the one where the huge trout lived. I was going to find that lake. I made a mental note that the riders had come in from the south.

Then there was a warm delicious feeling in my heart, and I thought of Liz, and of meeting her at the lodge.

After supper in the bunkhouse, while the other three guys stuffed themselves with meat loaf, baked potatoes and pinto beans, I washed up in the basin behind the back door, climbed into a clean pair of jeans, pulled on a shirt and found a damp rag to wipe the dust off my jump boots.

"What in the hell are you doing, Fearless?" Larry asked. "Washing up this early in the day could give you a case of mountain fever." Whatever that was.

"You're acting like you have a date," Bob Rakozy added.

"Maybe I do," I said. "With a little luck." They watched me walk to the road and turn into the trees in the direction of the lodge.

She was waiting just inside the door when I knocked.

"Hello," she said. She was even lovelier than the night before, wearing a short khaki skirt and a white sleeveless cotton top. She radiated health and summer days.

"Good evening," I said, and smiled, and we both looked at each other for a moment before saying another word. She broke the spell.

"I would like you to meet my parents."

"I'd like to, very much."

Mr. and Mrs. Holden lived behind the lodge in comfortable surroundings. They were older than I had expected, kind and polite. Mr. Holden managed the lodge and assisted the guides and outfitters during the fishing and hunting seasons. Mrs. Holden was the cook, and a terrific cook she was. In the next three days, I sampled her beef stew, salads, and blueberry pies.

"My parents run the place. I'm just here to get in the way," Liz said.

"Nonsense, we'd like to have her here with us for the rest of the summer," Mrs. Holden said.

"Three weeks, then I'm off to Boise," Liz said. "Now that I've graduated, I have to get a job. There's an opening for an assistant x-ray technician at Saint Luke's Hospital."

I told them where I was from, that I was studying forestry, and that I liked working for the Forest Service. They said they lived in southern Idaho but spent as much time in the backcountry as they could. "The mountains get in your blood," Mrs. Holden said. "Once you find this place, it's hard to leave, even for the winter."

"Well now," Liz interrupted, "later on, you three can talk all you want about the mountains. Bud's only going to be here three weeks, and I want to show him around."

"That shouldn't take much time." Mr. Holden winked at me. "The place isn't all that big."

I said goodbye to her parents, and Liz and I went out to the porch. Two hours of daylight remained. We walked along the road, past the horse pasture to the edge of the woods, then to the ranger station, down the runway to the creek and back to the lodge.

Sitting on the swing, we spoke of what we wanted to do with our lives. She wanted to get into nursing or become an airline stewardess. I wanted to study forestry, or at least stay a smokejumper for the summers.

We were still on the porch swing when darkness fell. In what seemed like only a moment, the lights went out at the ranger's cabin, then the warehouse and finally in the lodge itself. We listened to the night sounds. I took her hand and told her what a terrific evening it was. She kissed me on the check. "Tomorrow night?" I asked.

"Of course," she said, "and the night after that. You'll have to sample all of my mom's cooking."

And so it was. Each evening we hiked through the pines or walked to the creek to watch the salmon swimming upstream over the gravel bars. We took a shortcut through the pasture and rubbed the horses' noses, and when the sun dropped behind the peaks and darkness came quickly, we ended each day on the front-porch swing where we sat for hours, talking and laughing and telling each other of our adventures. Flying was one of our mutual passions. Liz said she wanted to be a pilot someday.

On Friday evening, I asked her if she had seen the big trout Mr. Kooch and Ted Koskella brought back from the high lakes. "No," she said, "but I heard about those fish. Those two go into the mountains every summer and come back with huge cutthroats and goldens, and they don't tell a soul where they catch them."

"There are a lot of lakes up there," I said, "high in the basins, hidden below the cliffs and rock slides. I'm going to get a map and figure out how to get to one of them. Would you like to go with me?"

"I'd love it," she said. I looked at her carefully: a beautiful girl who would venture into the mountains with me, high up in the peaks, to search for secret lakes full of monstrous trout. Back home, I'd never met a woman who wanted to explore the mountains, even if I'd suggested such a thing.

The Forest Service map on the wall was the logical place to start. We decided to begin our search with Lost Lake, about twelve miles up on the headwaters of Lost Creek.

"Tomorrow is Saturday, so we can get two horses from Bob Dustman," she said. "He told me we can use them any time we wanted."

"Let's do it," I said. She amazed me more each time I was with her.

In the morning, I picked up Liz at the lodge and we walked to the corrals where Bob Dustman waited. He was the assistant ranger—or the alternate as they were sometimes called—and worked for the district in the summers. During the other eight months of the year, he was on the faculty at the University of Utah Medical School.

He had caught a red-brown sorrel gelding and a paint horse

from the string that was spread out in the pasture. "These two are the best mountain horses we have. Just give them a chance to catch their wind when you're climbing. Be careful, and I'll see you back here tonight."

We swung into the saddles and Bob helped us make some adjustments to the stirrups. Liz waved to Bob, and we clucked to the horses and took the road south to where the trail started up Lost Creek.

It was dark and cool in the timber. The horse trail followed the creek, and as we rode, we could see the pines rising on the walls of the canyon. Clouds from the morning still clung to the tops of the peaks, where we were heading, but they'd disappear into the blue sky as the day warmed.

Riding on the trail was easy, the horses strong and eager to be moving. I led the way, walking the horse gently and getting accustomed to his gait. I turned in the saddle and looked at Liz. She rode well, erect and at ease on the sorrel.

When a creek fork came in from the left, we walked the horses around some alders, ducked under some limbs, and followed the branch up the side canyon. An open slope was ahead of us. I noticed how easily Liz moved her horse through the thickets at the creek.

Above us, the granite peaks beckoned. We walked the horses in the clearings, avoiding the thick stands of pine. Twice we saw brown flashes of mule deer bounding away in the trees and once, high on the side of the canyon, a herd of elk feeding on the grass below the rockslides. In three hours we had gained several thousand feet of elevation. The main branch of Lost Creek and the wooded valley fell away far below us. I followed Liz now, watching her move in her tight denims. I admired the way she rode, balanced and easy, while I had to work on keeping my jump boots in the stirrups when the paint crow-hopped over gravelly washes.

There must be a high pass up there in the rocks above us, leading into the next basin, I thought. A hawk circled in the sky, catching the rising currents of wind, and pine squirrels chirped in the trees as we rode by. The smell of the resin of the firs was euphoric. At mid-morning, we stopped to drink and water the horses. A spring bubbled from the rocks, cold as the snow chutes above.

"This water is the elixir of the gods," I said, trying to impress Liz. I did pushups at the edge of the pool, dropping and drinking the water, the way Seymour Peterson did in the movie *Telephone Creek*.

"I'll have to learn to do that too, since we don't have a cup," she said, laughing. The horses drank in the brook below us.

"The lake is up there somewhere," I said, looking up the canyon. "The walls of the mountains are still higher than we are. Looks like this valley levels off above us into a high basin, but I'll bet it's a false summit."

"What's that?" she asked.

"It's when there's a rise just in front of you and you're tired and you just know the top is there, but when you get there, it isn't, and there's another climb ahead of you, and you think that has got to be the summit."

"The lake could be up there anywhere."

"We could ride past it if we aren't careful," I said. "I haven't seen a horse track anywhere. Jack and Ted didn't come this way. But there are dozens of lakes up in these peaks."

"Guess that's why they call it Lost Lake," she said.

"It could take us several trips to locate it," I said, turning toward her. She loosely held the reins of the horses, patting their necks.

"Yes, it just might," she said. She didn't look like she minded the idea.

We swung back on the horses and climbed higher. The horses picked their way through the blowdowns and the rocks. As we climbed, the breezes were cooler and the leaves of the aspens rustled in the midday sun, as if the trees themselves were coming alive with our intrusion into their country, their mountains. And the aspens whispered to us—they had stories to tell, if we would stay and listen.

I watched Liz ahead of me. The sorrel was sleek and strong and sure-footed in this rough country. I thought of the Indian people who once lived here, roaming through these mountains, riding across the high passes from one green valley to another. Their time was less than a hundred years ago, and they had done for centuries what Liz and I were doing now. The game trails would have been here, the avalanche chutes, the aspen groves, they would all have been here, all the same, then as now. How great it would have been, how great it would be, if we could roam those mountains together, hunting the deer and elk, swimming in the lakes, laughing and making love in the sun.

A pine squirrel barked from atop a weathered log, breaking my thoughts. The paint horse nickered back. We came to the edge of a grassland that was dotted with boulders from a rockslide years ago. The white stone, imbedded with mica crystals, sparkled in the noon sun.

"It's so bright up here, it almost hurts my eyes," I said.

"Look." Liz pointed. "That line of trees over there could be the lake."

The horses sensed the moment and moved out quickly. On the other edge of the trees was the emerald water, hidden on the near side by a line of firs, and on the far side, by a slope of broken rocks.

We dismounted and took in the breathtaking beauty of the wilderness lake. A ripple from a rising trout spread out, making echoes of itself across the surface. We stretched for a moment, then sat together on the grass.

"Isn't this wonderful," she said.

"Yes. And not a sign of horse tracks anywhere, or old campfires, or footprints."

"Do you think we're the first people to ever see this lake?" she asked.

"I think the Indians were here years and years before us," I answered and put my arm around her waist.

"Mom packed us a lunch," Liz said, giving my hand a squeeze, moving away, and untying a pack from behind the saddle. She spread the food over a tablecloth on the grass. Mrs. Holden had prepared fried chicken, hard-boiled eggs, pickles, sourdough rolls and cookies. There was a thermos of iced tea and two tin cups, which I filled.

"To you, and to many more days at Big Creek," I said, raising my mug. She lifted her cup to touch mine.

"Thank you. That was very nice."

"Looks like your mother packed us enough food for three

days." *I'd love to spend three days up here with Liz.*

"The sun's warm and I feel pretty dusty," I said.

"So do I."

"Let's go for a swim."

"I'll bet it's cold."

"Cool, but not cold. Even though we're high up in the mountains, the sun is on the water most of the day." It was a big fib, and she knew it.

She got up and swirled her hand in the water at the edge of the grass. "You're right, warmer than I thought it would be," she said. "Anyway, my mother always told me to wait an hour after eating. And to watch out for guys like you."

"So did mine—about the swimming. Did you always wait?"

"Not always."

"And you never got a cramp, did you?"

"No, never."

"The water is crystal-clear. I can see to the bottom, and it must be twenty feet deep off those rocks."

"But I didn't bring a bathing suit, and I bet you didn't either," she said.

"We could swim in our undies—they're almost the same," I answered.

"For some reason I thought you might say that," she said.

"I'll dive off the bank and swim out to the center of the lake. We can come out on that sandy shoal."

"All right, but you go in first. And turn your back," she said. I noticed she was smiling, and wasn't turning hers.

I took off my boots, jeans and shirt, dove from the grassy bank into the water, did a slow crawl fifty feet toward the center of the lake, stopped and shook the water from my face. When I turned, she had taken off her boots and blouse and was stepping out of her jeans. She stood there for only a moment in her bra and panties and did a graceful arcing dive into the water. With a few strokes, she was next to me.

"You peeked," she said.

"Only for a second."

We swam and dove and talked, coming out once on the far sandy shoal to catch our breaths. The sun arced its way through the afternoon

Later, she put her hands on my shoulders and wrapped her legs around my waist and gave me a squeeze in the middle. I held her in the water to keep us both from sinking and then I kissed her on the lips, and she kissed me back. We swam together for a few more minutes, then both breathing hard, swam to the sandy bank and walked out. Her panties and bra clung to her tanned skin.

We lay back on the grass, drying off. Once in awhile, a shadow from a flying cloud raced across the meadow. The horses stayed in the shade in the trees, cropping the grass. Liz turned and faced the lake, her arms around her knees. "I could do this every day."

"If we don't find the lake with the big trout, we can ride this high country until we do," I said.

"It could take us ten years," she said.

"It could," I replied.

"It could take us forever." She looked over the water, her arms around her knees. Yellow butterflies bounced over the grass, and the afternoon wind sparkled on the water. The smell of the pines drifted over the lake.

"Larry Clark said that sidehill gougers lurk up here in the mountains," I said.

"Which one of your friends is Larry?"

"The guy with the blond crewcut and the muscles."

"The tall good-looking one?"

"That's him."

"Sidehill gougers?" she asked.

"Yes, and he said they're furry and they bite."

"I'm not worried."

"I better get closer. Just in case they come around." We were both dry from the sun. The touch of a breeze ruffled her hair. She turned her face toward me and I kissed her.

Shadows were setting up on the cutbanks of the canyon. The sun had dipped below the peaks. "We should be getting back," she said.

"You're right—we'd better be out of these mountains before nightfall."

"Until next weekend," she said, and she kissed me again.

"Next weekend."

Liz put her boot in the stirrup and stepped up into the saddle. Her hand circled her waist, sliding her blouse under her jeans. Then she sat her horse and faced me, framed by the late afternoon sky, and I would remember her like that for a long time. For a moment I saw again her long bare legs and her arms around me, laughing and talking in the grass. She smiled at me as if she knew what I was thinking.

"I already heard about the sidehill gougers," she said.

"I thought maybe you did."

"It was a nice try," she said, touching the reins to the gelding's neck. I watched her trot the horse downhill toward the open slope below, the wind tossing her hair.

The valley was in darkness when we rode up the path to the Big Creek corrals. Bob Dustman was waiting for us. "Thought maybe you got lost up there," he said, reaching for the reins.

"We rode all the way to Lost Lake," I said, swinging stiffly down from the saddle.

Liz dismounted and helped Bob remove the saddle from the horse. "You guys are going out tomorrow," Bob said.

"What?" Reality hit me hard.

"Half the jumpers at McCall went out on fires today. They want you back by tomorrow night. Rosselli will give you the details."

Bob Dustman took the tack into the stables, then fed the horses. I helped him, thanked him and said I liked working with him and hoped we would be back again this summer.

I walked with Liz to the lodge. Neither of us said much, each thinking that three weeks had suddenly been shortened to one, and we didn't want it to be over.

"Write when you get back to town," she said.

"I will," I said. There was a sudden hollow feeling in my chest. This wasn't going to be the end. I would see her again.

The moon was up and an owl hooted behind us in the timber. She took my hand and I kissed her, said goodbye and walked down the road and into the trees.

The next morning, the guys quizzed me about my activities the night before, but I wouldn't talk. At six a.m. we drove past the ranger station, the corrals and the lodge. The little settlement of Big Creek was quiet. I guessed that Liz was still asleep and wondered what she was dreaming about. I hoped I was in her dreams, and I knew she'd be in mine. The ride back across Elk Summit and down the South Fork took eight hours and before nightfall the four of us had jumped on a fire on Cayuse Creek.

Bill Yensen dipped his brush into a five-gallon can of gray paint and made a less-than-delicate stroke on the side of the Forest Service warehouse. Yensen was twenty years old, a little short of six feet tall, 170 pounds, as tough as they come and full of wit and conversation, no matter what the occasion. This was his first year and he had made it through the training. Bill would make smokejumping a career.

"Battleship gray, isn't it?" Bill asked.

"Gray or green, official U.S. Government colors."

"You know what they say in the Navy?" Bill asked.

"No, what do they say in the Navy?"

"If it moves, salute it, and if it doesn't, paint it."

"This fire cache building isn't moving," I said.

"We may get out on a fire in a day or two, the way these storms are moving over the mountains."

"I hope so. How many fire jumps have you made this season?" I asked.

"Eight. It's been busy."

"I'm up to seven, and want to get two or three more before school starts."

"Back to forestry?" Bill asked.

"Yep. It's in my blood."

"I'm going to stay until the season's over, clear into October. Maybe try to get on full time."

Bill and I had just returned from a fire at the headwaters of Goat Creek, up around the Bighorn Crag country of the Salmon

National Forest. Bill peeled off his T-shirt as the afternoon got warmer, displaying long red scratches down his chest.

"Looks like you were wrestling a grizzly bear, and the bear won."

"I told you I kept sliding out of the big ponderosa."

"You usually get scratches like that from those wild women you run around with."

"Don't I wish," Bill murmured.

"I did hear you yelp a couple of times up there on the ridge," I said, "while I was down there on the edge of the cliff, thinking my hip was busted. Why in the hell did you land in the biggest tree in the forest?"

"It seemed to be the safest thing to do at the time," Bill said, taking another stroke with the brush. "But they dropped me the pole spurs, with the short hooks, instead of tree spurs, and I'd skid off that thick bark every time I climbed a foot."

The story behind the conversation had started four days ago, when Bill and I had received the fire call. Reid Jackson was the spotter, and Jim Larkin flew the Travelaire. We took a route up the Middle Fork of the Salmon to the Goat Creek drainage, and spotted the smoke high on the breaks in some old-growth pine country. Below the rim, the canyon dropped off to steep slopes and cliffs. We circled the smoke just as the sun dropped below the horizon.

"There's the jump spot," Reid shouted above the roar of the engine, "with a stiff wind coming up from the river." I didn't like the looks of things. The jump spot was an open clearing on the rim above the cliffs, just below some huge ponderosa pines. Reid threw out the drift chute and three pairs of anxious eyes watched it descend. The chute landed way uphill in the trees; a pretty stout wind was blowing up over the cliffs from the river.

"Yensen, you're up!" Reid yelled, turning from the open door. Daylight was fading fast and I got the feeling we were pushing things. I didn't like the looks of those cliffs and the river and the rapids 2,000 feet below. I checked Bill's harness straps and gave his static line a last-second safety tug. "Go!" Reid shouted, gave Bill a slap on the back of his chute, and he disappeared out of the door. Reid and I filled the empty doorway to watch Yensen. Jim Larkin applied more power to the engine for the next pass.

What was Yensen doing? He missed the wide-open clearing on the slope by fifty yards and headed into the tall trees. Then there he was, snagged at the top of a huge ponderosa pine.

I began to wish I was home, reading a book or having a beer with friends; instead here I was, doing something I loved, but still it was something that could bust me up for the summer.

And this jump wasn't fitting together very well: the cliffs, the open slope, the wind, the upcoming darkness, and there was Yensen hung up in the top of that pine. Who could tell if he could get down. There had to have been one hell of a wind coming up from the river. I figured that Reid had reached the same conclusion, and I saw him hand-signal the pilot to fly the jump pass at right angles to the river, and into the wind.

When the wings of the Travelaire leveled, I scrunched down in the door. Reid watched the cliffs. We were on the jump pass. The Wright-Whirlwind engine kept droning on. *He's taking me farther out than Yensen. Not too far out over the canyon, Reid. The wind's not that strong. I won't be able to plane the chute back to the top of the cliffs. Make another pass. Not too far, Reid. Not too far!* The white rapids of the river were a half-mile below me. I would never question Reid's spotting, but just this once I was going to yell to him that we were out too far. Then the slap came hard. I stepped out.

Wow...too late.

"Damn it," I grunted, as the canopy popped and jerked me in the dimming light. I was way too far from the clearing. Yensen had taken to the trees deliberately. There was a wind, but not as strong as we thought. The ground above the cliffs was steep, much steeper than it looked from the plane and covered with boulders, and that's why Wild Bill Yensen took to the trees.

I turned the chute to face the cliffs. There were no options left—the parachute had to be planed with the wind. I reached up and grabbed both risers and pulled them down. During training we did twenty chin-ups on a monkey bar, and then ten more with one arm, if we were really tough. We could climb a twenty-five-foot rope, hand-over-hand, wearing six-pound logging boots, without using our feet. Now it would all come together in seconds, by reflex, to survive, so that when we did make it, we could talk and laugh about it later in a bar, over beers. And if we didn't, what the hell, it didn't matter anyway. Our mothers and our girl friends wouldn't know what our last thoughts were. But our jump buddies would—they'd know.

Pull the risers to your ankles. That was almost impossible. *Then pull the risers down to your waist and hold them there.* That was equivalent to a hundred chin-ups. But the adrenaline was flowing, and just plain fear made it happen. The squad leaders told us there would come a time when we'd have to hold the risers down and plane the parachute for the entire ninety-second descent, and that was why you climb ropes, they said, and why you'll do another fifty push-ups. And that time was now.

I drove the parachute forward, running with the wind. I had to clear the cliffs. The risers were down as far as I could get them, and my arms were locked like iron clamps. Now the options were really used up. The impact would be with the wind, into the side of the slope above the rock cliffs. There was no room to turn back into the wind and flare. The crash would be head-on.

The front of the parachute was down, air spilling out the back. Planing the chute added twenty miles per hour to the ground speed, and if the wind was moving at another twenty, the crash would be ugly. The top of the cliff went under my feet in a blur. I hit hard , and everything went black.

I don't know how long I lay there, but when my eyes opened, I was looking straight up and my right hip hurt like hell. Was it broken? I didn't think so. I could move my legs, but the trees on the ridge above swirled and blurred. The plane had banked and dropped in altitude as if to say it was sorry I had hit so hard and hoped everything was okay. I lay there watching the plane until it came around again, this time at a lower altitude. You'll be sorry you made me jump, I thought, and I'm not going to stand up until I'm sure I can.

What a sorry mess. What kind of a baby was I, anyway? There was Yensen up in that tree, and who knows what condition he was in, and here I was, not moving, and Reid and Jim in the plane worried about us both. And the smoke from the fire just visible past the rim of the canyon, the night coming on.

But this is enough. I'm alive, my hip hurts but I can move my legs, and the trees have stopped swirling. Time to try to get up. The plane made a tighter and tighter bank, lower and lower, just over me.

I stood and the mountains in the distance swayed back and forth. At least I could stand, so I spread the orange streamer on the ground. I unsnapped the jump jacket and it dropped at my feet. That felt better, getting rid of the harness and the extra weight. The terrain came into focus. The Travelaire acknowledged my orange streamer, Larkin waggled the wings, then made the cargo drop. Since Yensen's chute was hung up in the tallest tree on the mountaintop, the plane came in low for another pass, and Reid kicked out a small red chute attached to a pair of climbing spurs.

I'd leave the gear for now and take the climbing spurs up to Bill, the famous pole spurs with the short hooks, which he would still remember thirty years later. Darkness was not far off. My hip was bruised, and I favored it climbing up through the grass and rocks, but I'd go up there and help Wild Bill Yensen.

Bill stood at the base of the tree when I got there. "I used every foot of that hundred foot let-down rope to get out of that tree," he said through his big toothy smile.

"But you made it down okay."

"I was watching you through the branches. Didn't think you'd ever get out of that plane. You all right?"

"All right, but I hit the ground like a goddamn truck."

"It was a steep, rocky son-of-a-bitch," Yensen said. "Tomorrow, I'll take those spurs, climb back up that big mother and retrieve the chute." And that is when Wild Bill got all those scratches.

The fire turned out to be an easy one. Bill and I met the packer the next day and rode horses out to the river. We hitched a ride to Salmon in the back of a jeep, and the Travelaire came into the airport and took us back to McCall. Now, here we were, painting the warehouse, and Yensen kept looking down, admiring the long red scrapes on his chest and stomach.

A green Forest Service pickup truck headed our way. "It's Bob Fogg," Bill said. "I wonder what he wants?"

Bob pulled up alongside the warehouse. "Bud, I'm going into Big Creek after supper tonight to pick up a couple of jumpers. Want to ride as co-pilot? Rumor is you're sweet on a young lady back there."

"Sure, I'll go with you," I said. "Thanks for asking." I tried to look relaxed.

"I'll see you in the mess hall," Fogg said. "I'm eating there tonight."

The pickup went over the hill in a cloud of dust. I dropped the brush in the paint bucket. "I'm going to run over to dispatch and try to call Big Creek. Be back in five minutes." I wiped the paint off my hands and sprinted for the headquarters building.

Slim Vassar manned the radio, which he did every day of the week during fire season, ten to twelve hours a day. "Sure, I'll call Big Creek, and they can patch you through to the lodge," Slim said.

Mrs. Holden answered. "This is Bud. Is Liz there?" The radio crackled.

"She left an hour ago with her dad for Profile Summit."

"Bob Fogg and I are flying into Big Creek, and we'll be there in two hours. I was hoping to see her at the airfield for a few minutes."

"I'll tell her. She might be back in time," Mrs. Holden said.

I thanked her and replaced the radio mike.

Evening and the early morning hours were the only times to fly into Big Creek. The two jumpers we were picking up may have been waiting there all day, but the winds and the high altitude of the field made it too hazardous to land at mid-day, even with the powerful engine of the Travelaire.

At 5:45 the plane lifted off at McCall. Bob Fogg, the cautious pilot, left nothing to chance. He switched his eyes periodically from outside the plane to scan the instruments. I knew there were few places to set a plane down in those mountains if the engine should stop—it was all rocks and timber for hundreds of miles. Bob must have been reading my mind, and he explained how the magnetos worked, supplying sparks to the cylinders, even without a battery for electrical power. He said the plane could glide for miles without engine power and that would give him time to look for a field, a meadow or a thicket of jack pine. The Travelaire was designed to land at a slow speed, he said, close to fifty miles an hour, and the impact would be less than that of a speeding car. Still, there were many times I would have jumped rather than land on some of the postage-sized airfields the pilots used. "You're going to land *where*?" the jumpers often asked.

In an hour, we approached the Big Creek valley. The little airfield, nestled between the mountain on one side and the hog-back ridge on the other, looked like a green matchstick from four miles out and three thousand feet up. When we were at Big Creek three weeks before, we had driven in, and the field looked long and grassy when we stood next to it on the ground. And now, flying just below the clouds, I wondered how anyone could land on that little green pasture. As if to answer me, Bob Fogg pulled back on the throttle to begin the approach.

I prayed that Liz would be there. Maybe I was also praying we wouldn't crash.

The plane touched down from the north, bumped a foot up over the sod, settled down, then rolled toward the opposite end of the field where the familiar Forest Service buildings were located. The nose of the plane rode high, shaking as we taxied, and I couldn't see over it to look for Liz. Bob applied one brake, turned into the wind, and then I saw her, leaning against the fence and waiting to see where the plane was going to stop. One leg was crossed slightly in front of the other, and she smiled, her chin tipped up as if she had a secret, and just might tell you if you teased her just right.

Bob cut the power and the big prop shook and stopped. And then she came running toward us. I unfastened the strap from the open cabin door and was already out of the plane when she came up.

"You are the most awful guy in the world," she said. "It's been three weeks, and you've left this girl waiting, and only three letters from you!"

"I've been a little busy," I said. "We'd come in from one fire, repack our gear and go again."

"Maybe I'll forgive you—just this once."

I put my hand around her waist and we walked to the edge of the trees. I noticed two dirty-looking smokejumpers—Bob Donnelley and Pat Daly—at the other end of the fence line. They were picking up their fire packs and lugging them to the airplane.

"I jumped at the chance to come in, but we only have five minutes."

"I'm really glad you did. Dad and I just returned when Mom

said you called."

"Looks like this fire season is going to be a hot one. I'm sure we won't be back to Big Creek."

"I'll be in Boise in a week," she said. "I rented an apartment downtown with Syd, this girl I graduated with."

"Great. I'll call you in Boise."

"Come down. I'll show you around the city," she said with a sidelong glance and a wink. "And we'll dance at the Miramar."

"I love to dance... I learned a few steps in high school."

"And write as soon as you get back to McCall, even tonight." She stepped closer.

I glanced at the plane. Bob Fogg was going through his start-up procedures as Pat and Bob Donelley were stuffing their gear into the cabin. I wanted to kiss her, but it didn't seem right, here in front of the guys. It must be against the smokejumper's creed—or something.

"Climb aboard, Stud," Donnelley shouted, standing by the door. "Daly already has your co-pilot's seat."

Liz touched my arm. She was prettier each time I saw her. "I . . . I think I just might be in love with you," she said, and quickly kissed me.

My reply was lost in the roar of the engine. Grass rippled under the fuselage, and her hair blew back from her face. Donnelley was crouching in the door, waving me in.

I pressed a small white note in her hand, then turned and sprinted to the open door of the cabin. "It's...it's kind of a poem. It's for you," I shouted as the plane turned, the prop kicking up a shower of dry grass and pine needles.

The Travelaire made a steep climb out of Big Creek. At 1,000 feet, Bob banked the plane to the west. The valley below faded away in a gray-purple haze. Far behind and far below a young woman stood next to the fence line by the trees. She unfolded the note. I had read the words somewhere but I meant them for Liz.

I live but in your eyes,
the wish to win your smile
has lifted to the skies
this mortal for a while.

Bob Rakozy and Bud Filler

Chapter Ten

BOB RAKOZY AND THE LONELY LOOKOUT

In 1954, Bob Rakozy was my jump partner on two-man fires. Bob carried an Army Colt .45 caliber pistol on a web belt on his waist, and seldom was without it. He was stocky, with dark hair, a little reserved and on the quiet side. He started that summer and moved easily through the physical training and jumping requirements. On the line, he pulled more than his share of work.

Born in Michigan, he started parachuting with the Civil Air Patrol at sixteen. The minimum age for the CAP was eighteen, but he had lied to get in. When the North Koreans came across the 38th Parallel, he stretched the truth again and enlisted in the Air Force. In Korea, he rescued downed Allied airmen, often behind enemy lines. When he joined the smokejumpers after his return from the Korean War, he kept that .45 on his belt, or close at hand. The squad leaders didn't question him about it because they figured he had brought some ghosts back with him. And maybe he had. He was a hell of a good jump partner and we became close friends over the days we spent on the line together, and the weeks on project work.

❖ ❖ ❖

The smokejumpers had a way of communicating with each other at long distances in the mountains, by using the single word of "hoot." No one knows who started it, but by cupping the hands around the mouth and exhaling a great quantity of air the "hooooot" rolled along the ridges and echoed in the valleys. An answering hoot identified another jumper. We used short hoots when we landed in thick timber. A couple of hoots summoned the attention of firefighters on a fire, when only their silver hard hats could be seen in the smoke.

From time to time, ground crews used the hoot, but their volume and tone were inferior. Like herds of elk bugling to each other in the forest, the members of the jumping tribe could recognize each other by the sound.

The "hoot" and Bob Rakozy's .45 pistol came together one dark night on a ridge above the headwaters of Soldier Creek in the backcountry of the Salmon National Forest. A smoke was reported by the lookout on Short Creek Point at six o'clock in the evening. At eight, Bob and I jumped on a grass-covered hillside just down from the top of a timbered ridge. Smoke came through a low saddle and darkness was setting in when we dog-trotted, carrying a pulaski, a shovel and a tin of water, across the grass to the fire in the timber. The flames had already scorched a half an acre, but the fire stayed in the grass and the logs, and we had it under control by midnight.

The stars disappeared and clouds rolled in from the west, threatening rain, so we decided a tent was in order. Bob was ingenious in fixing things up in a hurry. Maybe it was his nature, or he had learned how to move fast to survive in Korea. We found a slender lodgepole, tied the top of the pole to the apex of one of the chutes, then fastened the edges of the nylon to rocks and in no time we had a teepee. The chute would keep off a mild rainstorm, or even snow, and would be none the worse for wear.

This had been an ideal jump: a smooth landing for each of us, the fire was out, and we had just finished a large can of spaghetti. We leaned back on logs next to the campfire. A cool night breeze came up from the canyon, so we had put on the high-collared jump jackets.

The sky was filled with clouds and the moon slipped somewhere behind the hills. I was watching the firelight on Bob's face, and his silhouette framed by the white teepee behind. The blackness of the night was complete and it came up to the edge of the campfire, a night so dark the faint line where the mountains ended and the sky began had disappeared.

It was after one o'clock when Bob reached across the fire, grabbed the steaming black tin of coffee with his glove and poured us another cup.

"This is the life," I said. "And this is the best part, camped back here in the primitive area, and not a person for miles, maybe hundreds of miles."

"Just the lookouts," Bob said, "and not many of them." He had unholstered the .45 and wiped the barrel with his handkerchief.

"You kinda like that pistol, don't you?"

"Yep, carried it in Korea. Used it a couple of times, too."

"Did you get them?"

"Probably. I didn't take the time to look. We were moving fast."

"It's isn't likely you'll ever need it up here in the mountains."

"There was a time I could have used it on a doctor," Bob said, his dark eyes concentrating on the dying embers.

"On a doctor?" I said. There was a story here but I wasn't sure I wanted to hear it.

"When I was overseas, my mother died in childbirth. I think the doctor could have saved her. Not long after, I came home." Bob continued to stare into the dying campfire. A burning piece of wood collapsed on itself, sending sparks into the sky.

"Sorry to hear that," I said.

"I don't like them even now, doctors that is," he said, his tone level and flat.

"I thought you said you were thinking of going to med school."

"If I can afford it. I plan to jump for a couple of summers yet."

"Then you better like doctors."

"I'm going to be different. I'm going to be careful."

"Well, you've got a long way to go, what with college and med school, but my guess is that you'll make it." We poured the dregs of the coffee on the campfire and crawled into our sleeping bags. The world of college seemed a long way off.

There is a moment in the darkest part of the night when the evening sounds cease. This is the haunting hour, the time long after midnight and before the first light of dawn, when the nocturnal hunters of the forest—the wolves, the cats, and the owls—have returned to their thickets. On the edges of the forest, the deer and the elk have been awake, listening, but the moon, not yet risen, leaves the night in darkness, and they remain hidden. This is the most magical time in the mountains, and strange things can happen at this hour.

At three o'clock I woke to the sound of a long wavering howl below the saddle of the ridge. A few moments later another howl echoed below the camp.

"What in the hell was that?" Bob sat up, spring-trigger alert from his combat days.

"I don't know. I heard it too." I unzipped my sleeping bag, and sat upright inside the white teepee.

"Listen!" Bob said.

Again the long eerie howl came up from the meadow. Only this time, whatever was howling was closer.

"I think it's a wolf," Bob said.

"A wuf? There could be several of them. The wufs, that is," I said, feeling secure inside the teepee. "We're surrounded," I said. "Anyway, I'm not worried, you've got the gun."

"Maybe it's Bigfoot," he said. I remembered Roselli's comments about Bigfoot the day we ran head-on into the goats.

"Nobody has actually seen Bigfoot," I said. "We might be the first." I was looking for a match to light; we had no flashlights. I could see Bob's silhouette against the last of the flickering campfire on the other side of the nylon. He was sliding the .45 out of its holster. Our imaginations were on overtime.

"I think they're wolves."

We threw back the loose nylon of the chute and stepped into the darkness. The strange wavering sound came up from the saddle and floated again over the meadow.

"There's something in the trees. There!" Bob said and pointed. "I saw it move." He pulled back the slide of the .45. A cartridge being positioned in the chamber and the slide going forward made a sharp metallic "clang" in the night air.

"Hang on, Bob," I said. "We've got to find out what it is." Then I remembered that the fire dispatcher had told me the lookout on Short Creek Point, who had reported the smoke, said he might come over to help. From the air, the lookout tower seemed more than ten miles away and over rough terrain, and I'd forgotten him until then.

"Just being ready," he said.

The wavering howl came again on the night air. This time the sound was further away. Bob lowered the .45.

"It's the guy from the lookout." I was fully awake. "He's trying to locate us."

"Let's hoot him over here," Bob said. With that we both let out long hoooots, the noise piercing the wilderness night. In a minute, a dark figure emerged from the treeline and ran toward us across the meadow.

"Hey guys! I'm from Short Creek Point. I'm here to help," he said, breathless. In the darkness we couldn't see what he looked like, but he ran up and threw his arms around Bob, then turned and gave me the same bear hug.

"Ron Buckbee," he said in a high voice. Now he was almost shouting. "But just call me Buck."

Ron Buckbee was nineteen, wiry, energetic, slinking around like a nervous cat. He was in a state of delirium. "I haven't seen a living soul in months," he said, still out of breath. "Yes sir, you guys are a sight for sore eyes. The Forest Service drops my food from a plane into the meadow. The packer rode in two months ago with mules but he only stayed long enough for a drink of water. Tomorrow, tomorrow, you can come back with me to the lookout. Come on back with me."

We finally got the chance to talk and introduced ourselves.

"What kind of a noise was that you were making?" Bob asked. "I almost shot you."

"I was making a hoot," Buck answered. "Like you jumpers make."

"That howl sounded like a bear with a sore butt," I said.

"Or a bull moose in rut," Bob added. "Which could be

worse."

"I'll go to the next fire and practice," Buck said, bouncing back and forth between us.

"Actually, we wouldn't have shot you," I said, hoping he'd settle down.

"Unless you'd have kissed us," Bob said. "Listen, the fire is out. No need to help with that. Let's go back to bed. Buck, you can stay with us in the teepee. Just roll up in the other chute. Grab yourself a can of that orange juice over there."

Bob Rakozy would make a great doctor some day, I thought.

Buck needed no urging. "And tomorrow you can go back with me to the tower," he said. "It's only a few miles."

In the morning, Buck repeated his offer to visit his lookout home, only this time it was more of a plea than a request. Bob and I looked at each other. "Why not," Bob said. "The fire is out—only a few smokes this morning. Let's go check it out."

After breakfast, the three of us set out cross-country for Buck's lookout. The trek took us along the main ridge of the mountain, a series of steep ups, and then steep downs. The narrow top resembled the saw-toothed spines of a dragon when I turned to view the route we had just traveled, to get a picture in my mind of our back-trail. Just in case. Who could tell where Buck was leading us.

We traveled more than ten miles before we finally came to the bottom of a sharp peak, rising as a spire of granite rock before us. Far above was a wooden tower, perched on the rocky top. Buck led the way, jogging up the switch-backed trail among the rocks. Bob and I sweated hard, panting, trying to keep up. And we thought we were in good shape.

"This guy is a mountain goat," Bob said, behind me. "How does he stay in shape, sitting all day in a fire tower?"

"When we get to the top, if we ever get there, let's ask him."

We finally reached his lofty home, and Buck was hardly out of breath. He said he climbed down from the lookout twice a day and brought back five gallons of water each time. He used it for drinking, for washing dishes and clothes, and for cleaning the many windows on the four sides of the tower. The inside of his lookout home was spotless.

"See that clump of aspen down at the treeline? I built a springhouse there." Buck pointed to a spot 500 vertical feet below us. "That's where the meat and vegetables are stored."

"You'll make someone a good wife someday," Bob said.

Buck couldn't stop talking. He told us of the animals he saw, of the marmots in the rocks who became his pets, of the hawks soaring on the afternoon thermals, of the storms that came in from the west and hammered his tower. He showed us the radio he used to call twice a day to the guard station. He said the Forest Service guard wouldn't talk much, just wanted to know if he was okay. When the plane came to drop food in the meadow, the pilot was eye-level with the lookout. He rocked the wings when Buck had mail, sending him scurrying to the meadow, but lately, he seldom waggled the wings. Buck showed us the two letters he'd received in three months. They were dog-eared and well read. A lost love? He didn't offer to tell us.

"Stay for supper," Buck demanded. We decided we would, even though we'd have to hike back to the ridge camp in the dark.

Buck took an hour to prepare supper while we sat on the catwalk around the tower, observing the distant mountains and looking for smokes. Finally he announced that dinner was served.

A well-scrubbed canvas pack manti was draped over the wooden table. In the center stood two lit candles, balanced delicately on short carved pine branches, and at each setting was a tin cup of a sweet juice. He said he had made the dark liquid from mountain berries. To us it tasted better than a fine French wine—not that we were connoisseurs of fine wines, or fine anything. The menu was steak, medium rare and perfectly prepared, mashed potatoes, gravy, corn, sourdough bread, huckleberry jam and a three-layer chocolate cake with white icing an inch thick, followed by steaming coffee in white porcelain cups.

"You're one hell of a cook, Buck," Bob said, finishing his second helping.

"This icing is terrific. How did you make it?" I asked, sounding like my mother on a visit to a friend's house for tea.

"Bacon grease," Ron Buckbee said. "It's good, isn't it?"

I looked at the cake and then at Bob. The cake and icing *were* excellent.

After supper was over, Buck showed us the equipment in the lookout. He demonstrated the alidade, the sighting device used to determine compass bearings. He showed us his firepack. He said lookout operators were expected to chase down smokes if they were close.

The sun, almost red now, touched the far peaks, and we said it was getting late and we had to get back and check the fire. We thanked him for the supper and wished him good luck.

"You'll write, won't you?" he asked and pressed small pieces of notepaper into each of our hands. "My address is on that. The ranger will see that I get your letters."

"We'll write," Bob said.

"Sure, we'll write," I added.

"Okay, but don't forget. It gets—it can get a little lonely up here." Buck looked serious, and I suspected I saw a tear roll down his check as we started down the trail through the rocks in the fading daylight.

"Now, don't forget," Buck said. He waved his arm.

"We'll write, Buck," I said and waved back.

"Yes we will. Now take care of yourself," Bob shouted as we dropped out of sight of the lookout in the lee of the boulders.

We moved fast back to camp. After we'd climbed one of the saw-toothed hogbacks, Bob paused, caught his breath, and said, "Buck is a heck of a good cook, but he can't hoot worth beans."

After a demanding two weeks on fires, Bob and I were drinking Lucky Lager at the Yacht Club saloon. "Let's write to Buck and thank him for his hospitality," I said. And we did. We composed a letter, trying to make it funny, stamped the envelope, then handed the letter to the pilot the next day, because we were suited up and ready to climb aboard the Travelaire, and the pilot was going back to town and we weren't. We never heard back from Buck.

For reasons unclear even to me, I often think of Ron Buckbee, usually after climbing a high peak and looking down at the forests below. I can see Buck standing on the porch of his lookout, binoculars in hand, searching the summer mountains for smokes, and waiting for letters that never came.

As for Bob Rakozy, he went to medical school and became a well-respected physician in Lake Oswego, Oregon. I often wondered if he still carried the .45.

Wayne Webb and Ray Beasley

CHAPTER ELEVEN

BARROOM BRAWLS AND THE JUMPING CAT

"If you guys averaged one malfunction out of four openings, you had to be tough—and you had to be dumb."

USFS parachute rigger, Missoula, Montana, 1995, citing old jump statistics.

Snow cornices still hung on the ridges in the backcountry; the Fourth of July was not far off. The fire season was slow in arriving and we wanted some action; we wanted to fight fires. A week or two of ninety-degree days and dry winds, and we'd be out there. But until the lightning storms arrived the brush needed to be piled, the most monotonous job in the world. We'd rather do almost anything but stack brush. We piled it along the roads in the summer and crews burned it in the fall, making the mountain roads in the forest more attractive to campers and tourists, and decreasing the chance of man-caused fires. The procedure was to chop up and remove the brush for fifty feet on each side of the road until the forest resembled a park. The brush came from logging slash, trees from blowdowns, and natural forest litter.

By the first week of July we'd cleared the sides of the road to Hazard Lake. The road across Lick Creek Summit was spotless. The forest highways near Krassel Ranger Station were as clean as a whistle. Our mothers would have been proud.

Larry Clark, a self-proclaimed expert at brush piling, said it was an art and not a science, and had to be done carefully. The boys built square piles, round piles and tapered piles. There were piles with pointed prows that resembled boats. Don Clark, Larry's sometimes jump partner and companion—not related, thankfully, they both said—was skilled in making criss-crossed stacks of brush. We worked hard at whatever job we took on, but by ten in the morning we were ready to call it a day, and there were still six hours to go.

Other projects kept the firefighters busy until the season got rolling. Bunkhouses were painted, the ranger offices were spruced up in white with green trim, and the jump tower was whitewashed. It was said that Len Walker, with a hangover after a wild night in town, fell asleep at nine in the morning on the top deck and remained there until noon, the paint brush dried in his outstretched hand.

Before the lightning storms moved in from the west, the jumper crews remained in town, except for small units of guys who were low on the ready roster, and who were dispatched to ranger districts on the Payette or Boise forests.

The jumpers had a reputation for creating parties and having fun anywhere we went. We'd accept any challenge, and we never backed away from a fight. On Friday and Saturday nights, we'd move in groups of two, three or four from the Chicken Roost to the Shore Lodge to the Yacht Club, and usually end the evening at the Foresters Club. There, at one in the morning, the shades were drawn on the door, and the fun continued until the early hours. The slot machines continued to clatter: gambling was legal in Idaho.

There were pretty girls everywhere in the summer in McCall—most of them coming up from the Boise Valley—and they liked to spend time with the jumpers. The smokejumpers for the most part were a love 'em and leave 'em bunch of guys who stayed away from relationships, since they'd face being in the mountains for the rest of the summer. Several of the good-looking local women said they would never date a smokejumper, because they heard the jumpers were dangerous. That was a shame, the jumpers responded, because the ladies didn't know what they were missing.

Besides having the basic woodsmen's skills, the first-year men needed to be strong and athletic, have exceptional endurance, and had to have thick skin. A guy could be a star football player, or a past member of the Army airborne, but if he lacked the ready smile after a prank was pulled on him his first year, he wouldn't make the grade, and some didn't.

Paperlegs Peterson was a master at insuring that the new jumpers had the mental toughness. If they could make it past Paperlegs, they'd be back year after year. With a Lucky Lager in one hand, feeding silver dollars into a slot machine with the other, Paperlegs would tell the new men gathered around him that the Pioneer chutes were proving faulty, and most of them weren't opening. He said there would be two or three of the new group who wouldn't complete the seven training jumps (which was true), and a memorial would be erected in their behalf (which was not true) at the McCall airport, near the meadow where they piled in. Paper's blue eyes would shine, and a logger seated at the bar would buy another round of drinks in memory of the new guys, and then they'd all laugh together, having heard the stories before.

He was teasing them, of course, but some of his stories were true—the Pioneer parachutes really did malfunction. Paperlegs told his eager audience about the dearly departed parachutist whose main chute streamered, the lines twisting and constricting the "silk" from ballooning, and the whole thing—a tangle of ropes and canopy—heading fast toward the ground. The parachutist pulled down hard on the risers, jerked the shroud lines, but to no avail, and he was dropping fast. He pulled the rip-cord on the reserve, but neglected to throw the silk into the wind, and it furled and fluttered and caught on the steel cage of his leather helmet and then swirled around the main chute, already a falling rocket of lines and nylon. Paper finished his beer and laughed. "The jumper ran out of ideas and air at the same time," he'd say.

A trick that was pulled on an older jumper—never on the new ones, they couldn't have handled it—was to slip one of the old original Eagle chutes, which opened like a shotgun blast, into the victim's locker. When called to a jump, the chutist would strap on the Eagle without looking at the serial number. The Eagles were thirty-six feet in diameter and were designed with a skirt around the perimeter. When the canopy opened, the jumper would be snapped within an inch of his life, and possibly lose future use of his entire spinal column. It was said that, on the opening of these old canopies, a parachutist's helmet could fly off if it wasn't secured, and sometimes even if it was. It was common to see cigarettes from shirt pockets fly away into the sky on the opening shock.

Gene Ellis, the tough squad leader from eastern Idaho, said they'd never get an Eagle chute on him. While he was ranting about the inhumane use of these ancient devices, Reid Jackson gave the nod to Spencer Miller, who pulled an Eagle from the locker and snapped it to Ellis's harness, all the while listening to him brag that no one in the outfit was smart enough to put one over on him. Reid Jackson, who spotted Ellis on the jump, said you could hear the crack of that chute opening for miles, almost as far as you could hear Ellis curse his jumpmates.

One of the tricks the old guys pulled on the newcomers was to fill the chute cover with confetti. When the canopy opened, the confetti scattered in the sky. The jumper would be alarmed, but only momentarily, because all the way to the ground he would be planning on how to get even.

Another stunt was to insert short pieces of shroud lines just inside the back cover of an unsuspecting jumper's chute. Inside the plane, usually just before the jump pass, the perpetrators would examinine the jumper for the final safety check, and "discover" the shroud lines coming out. The perpetrator would withdraw his combat knife and cut away the "extra" lines, stuff the loose ends back inside the canvas cover, and announce the chute was now "okay." The stunt was usually pulled on first-year guys who were smart alecks, or had big egos, and there were always a couple of candidates in each group who got the full treatment.

It wasn't as easy for the jumping cat. A big gray cat, owned by no one, who hung out at the parachute loft decided he wanted to jump, or so it was declared by the parachute riggers. The cat was the smokejumpers' mascot and head mouser. To acquiesce to the cat's request for this adventure, a personal cat harness was made with a small parachute attached to miniature risers. Early one Saturday morning the cat was fitted into its rigging and parachute and given a practice jump off the tower. Attended by eager hands, the feline was thrown out from the tower at a safe angle to insure it and the chute wouldn't collide with the timbers. The cat, as all creatures of its kind do, landed on all four paws, and because it didn't squeal, the spectators assumed the cat was ready for a real jump. So the entourage proceeded to the airport where a Cessna 172 and a pilot were waiting. The smokejumpers' mascot was about to become a Certified Combat Jumping Cat.

The plane gained an altitude of 1,000 feet over the meadow. Spectators on the ground had anxious eyes to the sky, for this was history in the making.

As the aircraft made a slow circle, from the open door a tiny gray ball appeared, and a moment later a pink and white parachute opened. A cheer went up from the crowd. The cat, all four paws pointing straight at the ground, descended in a slow swinging motion. But nature had not prepared the cat for this activity and soon it was obvious it didn't like the rapidly approaching meadow. The cat struggled in its harness for a moment, and then like all climbing creatures it went up the shroud lines like a shot. When it got to the nylon canopy there was nothing left but lines, puffs of a parachute, and a whirling gray ball. Fortunately this all happened close to the ground and the cat landed in a clump of sagebrush. A gasp went up from the spectators. In unison they ran to the assistance. The cat, now on its seventh or eight life, slipped from the tangled harness, shot like a cannonball from the sagebrush and ran into the woods, never to be seen again.

None of the smokejumpers went looking for trouble—the Forest Service frowned on it. On the other hand, none of us backed away from someone who got out of line, usually in a smoky bar. Few strangers messed around with the jumpers, because we had a reputation for being downright tough if we had to be. In the early fifties, the years of parachuting and development of small-unit firefighting techniques and expansion in aerial fire suppression, many of the men in our unit were combat veterans from World War II. Most of these guys were professional adventurers and had paid their dues in the air and on the battlefields, and they took no flak from anyone.

One such jumper was Bus Bertram, a tall, raw-boned, heavily

muscled woodsman and veteran tail-gunner on a B-17. Bus was blond, always kept the short military crew-cut, had taunting blue eyes, and weighed at the maximum 180 pounds. He had played football for Boise High School before the war, and had years of experience jumping and on the fireline. Bus was about ten years older than I, and set a good example, I thought— at least most of the time—either on a fire or in a bar. Bus looked after the new guys.

Bertram was fast with his fists. Ray Mansisidor, who jumped at McCall in the late forties, said Bertram was the fastest street fighter he had ever seen. Ray himself was a rough, tough Basque from Homedale, Idaho, who had honed his skills with the sheepherders and ranchers in the mountains of Owyhee County.

Ray told me later that he considered both Bertram and himself friendly but aggressive, but he thought that Bus didn't like him, and he knew that in the unit there was going to be a showdown someday between them to see who was really the "bull of the woods." Ray said that Bertram, who was five inches taller, approached him as he was stuffing equipment in his jump bag.

"You think you're tough, don't you, you big ugly Italian?" Bus said, walking up to him.

Ray said later, "I looked up at Bertram and he looked a foot taller than me, and he was glaring at me and he looked serious, and I was sure he wanted me to swing on him."

Ray said he finished tying down the top of his fire pack. Bertram continued to glare.

"Pretty tough, Bertram," Ray said, and positioned himself for a swing.

"Well, I like you, you tough little Italian," Bus said, relaxing that devilish gleam. Then he laughed and put his big arm around Ray's shoulder.

Ray said he told Bus he was Basque and not Italian, but "I did breathe a sigh of relief." He and Bus were the best of friends after that.

Ray's last year of jumping was in 1949, but he continued to make free falls when he could find a dusty parachute and locate a pilot who would take him up. From the lake-side of the Shore Lodge we watched him jump on Sunday afternoons. He'd go out of the plane just over the end of the boat docks and then wait until he was several hundred feet above the water before pulling his rip cord, making us gasp and giving us heart palpitations.

Bertram jumped until 1953. We became good friends; I liked to think Bus was watching over me, and teaching me about fires and anything else he wanted to tell me about. I had been on several large fires with Bertram, and his favorite saying seemed to be, "Let it crown, boys. We can't stop it now. It's going over the hill. We'll get it tonight when the winds are down and the temperature drops." And we did. That was how I learned that fighting wildfires at night was the best tactic.

One Saturday night I also learned how fast Bus was. Bertram, Paperlegs and I were downing some brews and exchanging jump stories at the bar in the Yacht Club, a downtown saloon and dance hall on Lake Street, just aft of the boat docks. The drinking, music and dancing were getting better and hotter as midnight approached. It was an evening when the spirits were flowing. None of us went out looking for a skirmish, if we could help it, but then sometimes we couldn't help it.

A couple in their mid-thirties came in and sat on the two empty bar stools between Bus and me. The woman—dark hair, sophisticated and attractive in a red dress—was having an exchange of words with her partner, a tall, mouthy slick-haired flatlander. He pushed her into me, she apologized, and that's when I noticed how attractive she was. A moment later she pushed her foul-mouthed companion into Bertram's shoulder, who hardly budged, but mentioned the two of them were interrupting the friendly atmosphere of the establishment. Then she came my way again with a shove from Big Ugly, and she let out a string of cuss words that would make a logger blush, and I figured she was getting annoyed. The ugly one with the slick hair then slapped her, hard, with a smack you could hear over the country music, and this time she bounced into me hard. Her head recoiled and she put a hand to her cheek. Now I really wanted to punch this guy out. I had boxed through high school and just wanted to tag him one.

"You can't do that to a lady," I said, and my fist came back for a swing around the woman, who was now sitting, somewhat stunned, in my lap, her hair in my face, her perfume overpowering.

"Oh yeah," Ugly said, looking for an opening, "I'll hit anybody I want to."

But that was when I learned how fast Bertram really was. Hardly changing his position on the bar stool, his right came around in a blur, hit the woman-slapper squarely on the jaw, lifted him off the stool, and catapulted him in a crash to the floor, almost to the opposite wall. Bus stood up, his steel-blue eyes glaring, walked over to him and said, "Now you just lay there and play dead." That line became a classic, repeated for summers to come.

The woman in the red dress regained her composure, and went to his assistance. "You didn't have to hit him *that* hard," she said. She took her companion's arm and they left.

Slightly offended in being outmaneuvered in what I thought at that time might have been fun, I said to Bus, "He was swinging at me. I would have gotten him."

"You were too slow, Fearless," Bus laughed. Paperlegs was talking to the bartender, acting like this was just part of the evening's festivities. Somebody ordered another round of Lucky Lagers.

Another smokejumper who had almost no physical equal, at least when it came to outright strength and endurance, was Ken "Moose" Salyer. Moose was an all-collegiate wrestler from Iowa State, and tipped the scales at more than 175 pounds. His neck size was about eighteen inches. Moose started jumping in 1954 and kept it up until his death in 1965. In his first year as a Ned, the other jumpers pulled pranks on him, like filling his elephant bag with river rocks on the first training hike, which made the pack weigh more than 150 pounds. Moose came in first anyway on the race from the practice jump to the base. He just grinned after stopping at the finish line and pulled the rocks from the bottom of the bag.

Moose had an easy-going smile for his jumper friends—and we were all his friends—and being the wrestler that he was, suffered no fools when one of us got behind in a barroom brawl.

Another night at the Yacht Club, while I was dancing with a brunette, a young tough came up to me, tapped me roughly on the shoulder, and told me not to dance with his girl. The young lady I was dancing with was a good friend, and I knew she wasn't going with this guy.

"She's not your girl, and I think she wants to dance with me," I said. I also told him to leave us alone, and then slipped the young lady behind me, because this guy was trouble, spelled with a capital T.

The tough took a swing, deflected, then another, blocked. It felt good to be boxing again. Now I really wanted to smash this turkey. But two hairy arms reached around me from behind and wrapped around my arms and chest. Now there were two of them.

Moose watched all this and set his beer gently down on the bar. I swung around so the first guy was between me and the bar, and Moose simply took one step forward, grabbed the first guy, and folded his arms across his chest. Then Moose gave him a bear hug he would never forget. That left me with just the guy behind me, who was still trying to squeeze the air out of me. I did the wrestler's shoulder drop on him, pulling him down in front, then hit him once below the ribs. He went as limp as a bag of potatoes. And then, just like in the movies, Moose and I grabbed each of them from behind and threw them out into the street.

"And stay out," Moose said.

Hairy Arms wiped his sides, cleaning off the dust. "We're coming back and kick your butts," he mumbled.

"Anytime," Moose said, turning back inside. He was already losing interest in the matter.

"We'll be here at midnight," the tough guy said, "in the alley."

"We'll be here. Just knock," Moose said, and closed the door.

"I hope they come. I'd like to pop that smart-ass a good one," I said.

"I doubt if they'll be back," Moose said.

I waited around until midnight, and I hoped Moose would too, because I accepted their challenge and told them we'd be waiting; and if they came back and we weren't there, it would go against the smokejumpers' creed of never backing away from a skirmish.

They never returned. The thought of facing Moose must have been too much for them. Anyway, the cute brunette I was dancing with seemed to enjoy the attention.

Years later, after I had left the jumpers and saw Paperlegs in town, he told me Moose was dead. There was a pause when I couldn't speak. I thought of that big toothy grin of Moose's, his we-don't-have-a-problem smile, the long hikes off the fires with him, and his not even raising a sweat. The fight in the bar went through my mind.

The official story came out that Moose Salyer and pilot Byron Knapp were killed on Norton Ridge in the Boise National Forest when their plane crashed making a cargo drop, after encountering severe air turbulence. It was on July 9 in the summer of 1965. Moose was one hell of a guy. I still think about him, and all the others who gave their lives in war and peacetime, working to save what they loved.

Pat Daly

Chapter Twelve

THE CURSE OF TRAPPER CREEK

Wildfires burned in the mountains. Every jumper was busy and the adrenaline ran high. Coming back from a fire, we'd replenish our fire packs, grab new chutes from the loft, take a quick shower and jump again, sometimes within hours. We were happy because we were working long hours on overtime, making some serious money, and doing what we had signed on to do.

A series of storms had crashed against the peaks. The fires that had been started by lightning on the ridges worked their way down into the valleys, and acre on acre of timber was being destroyed. All the available ground crews, forest guards and rangers were on the fires, and Indian crews from Montana were being flown into Idaho to build lines on the larger fires. These were the project fires and several were going at one time, one burning on Trapper Creek, near the Middle Fork of the Salmon River.

On a busy Saturday afternoon, Reid Jackson came into the bunkhouse and said that Lavelle Thompson wanted to see me. Lavelle was the fire control officer on the Payette and his word was law. He let the jumpers carry out their tasks day-to-day and rarely bothered us in operations, so this must be something different. I double-timed over the hill to his office.

Lavelle Thompson got quickly to the point. He was looking at the wall map of the Payette, with pins showing each fire. "We have another job for you," he said, gesturing to the map. "You're studying forestry and if you want to gain some experience in fire control and administration we need you at the Cold Meadows Guard Station to refuel the helicopters and help organize the Indian crews hiking into the Trapper Creek fire."

"I'd rather jump," I answered modestly, as Mr. Thompson was the big boss.

"We need you at Cold Meadows," Lavelle said, making this a direct order and cutting off any further discussion. "Hobbie Bonnet will be here at the supervisor's office, coming in off timber sales," he said. "Let him know what you need at Cold Meadows."

Lavelle looked up from his paperwork and glared at me. "Four new fires have been reported in the last hour. The big one is on Trapper Creek. Dub Horn is there at the guard cabin with the pack string. We'll fly in food and firepacks from McCall for the Indian crew. The gasoline will be for the chopper. The Indian crews will arrive tomorrow at noon—be sure they get fed and then move them out on the trail to Trapper Creek. Bob Fogg is flying supplies in there this evening in the Travelaire. Be with him. Take it from there," Lavelle said.

"Yes sir," I said. This was a new responsibility, and Lavelle said I'd learn more about Forest Service firefighting, administration and supply. But damn it, I'd rather be jumping with the guys, and out on the line with a shovel in my hand. That's where the real adventure was ... that's where I belonged.

Cold Meadows was a nice place to work, a beautiful high valley in the forest, far back in the Idaho primitive area. The Travelaire lifted off at six. We flew over the peaks west to Cold Meadows, circled the runway once, and then flew on and checked out the Trapper Creek fire from an altitude of 500 feet. Sixteen McCall jumpers were on the line as well as twenty ground crew personnel from the Krassel fire camp. The fire had continued to blow up and jumped the lines when the winds came up in the afternoon. Spot fires were starting on the adjacent ridges, and the spot fires were turning into major fires on their own. Holding down the line and chasing the spot fires was too much for the thirty-six-man crew. The fire was crowning, blowing up and winning the battle. Two crews, totaling sixty Indian firefighters, were coming in the next day. Bob banked the Travelaire and we returned to Cold Meadows.

The plane touched down fast but gently on the long grassy strip. We could see Dub's horses and mules in the pole corral just off the north end of the guard station. The low hills rolled away to the north and east, gradually giving way to the Middle Fork and Salmon River canyons, thirty miles by horse trails. Cottonwood Creek ran through the center of the meadow, curving its way past the log guard station and paralleling the airstrip. The creek was noted for its scores of pan-sized trout, easy to catch and great eating after they were fried on the wood stove.

Jim Larkin ferried food rations, sleeping bags, shovels, pulaskis and gasoline in the Ford the next morning. A Bell helicopter buzzed in and out of the strip, landing at the far end of the runway. I helped the pilot refuel. After filling the tank, he set two five-gallon cans of gasoline next to his seat, closed the door and lit a cigarette. The fumes of the gasoline were visible behind the Plexiglas canopy in the cockpit and I quickly stepped back and wondered why the damn thing didn't blow up.

Dub Horn led a sleek pinto gelding from the corral and brought him over while I was arranging the fire-fighting equipment. "Here's a horse to save you some time running back and forth to the field," Dub said. "His name is Moonshine." I said thanks and looked Moonshine in the eye. He stared right back at me. Yes, we'll get along, I told him. I threw on the hackamore and galloped out bareback.

The Indian crews began arriving at noon, ferried in by groups in the Ford. They had brought sack lunches with them. There were too many to assemble near the cabin, and I motioned them to wait off the edge of the meadow, in the pine-tree shade.

The Indians were an elite group of ground fighters. Most of them were square-jawed husky men who could work the fireline from dawn to dusk. They were happiest when the fire camp cook provided them with plenty of bread. On other fires I had seen them pass up seconds on steak and walk away from the grub line with a half-loaf of bread in each hand. It must have worked: they were tough on the line and commanded a lot of respect in the fire camps.

Walter Black Bear was one of the Indian foremen. "Good

fire up there," Walter Black Bear said, and nodded at me.

"It's a big one," I answered. "You've got a fourteen-mile hike along Cottonwood Creek to Trapper Creek. You'll be in there tonight."

"The rest of my crew's coming on this plane." Walter Black Bear looked pleased as he saw the Ford touching down a half mile away. "Full crew's here now. We'll go to work."

"You get enough for lunch?" I asked, wanting to be sure his crew was happy. "You'll get a food drop tonight at fire camp."

Black Bear's face glistened in the afternoon sun. "We ate enough. We'll get that fire knocked down pretty quick. Lots of fires in Idaho. We'll be here all summer," he said, probably thinking of the overtime pay. Walter Black Bear wore a red neckerchief. Many of the Indians wore colorful shirts, bright bandannas around their necks, or a feather attached to their hard hats.

The Indians were sitting on the grass, waiting for orders to move out, now that the last planeload had been ferried in from McCall.

I noticed a large Indian man, tall and heavily muscled, sitting alone just off the grass and outside the shady perimeter of the pines. Through the lunch hour he sat there, eating in solitude.

There was a rustle of hard hats, a wrinkling of paper-sack lunches, a stirring as the Indians prepared to move out along the trail. But the lone Indian remained there, sitting, cross-legged, his eyes off in the distance. Then suddenly I stopped. A half-dozen blackbirds flew above his head, and in ones and twos, they swooped down over him, missing his black braided hair by inches. Where did they come from and why were they flying over him?

"We'll go now," the leader said. I had moved away from the Indian man and the birds as if by some strange reflex. Strange, I had never noticed blackbirds here in the mountains before.

"One thing, Black Bear," I said, speaking quietly to the Indian leader. "Will he go?" I asked, nodding toward the lone man still sitting there.

"Dan Big Bull. He will go," Black Bear said.

"Why are the birds above him?" I asked.

"He is cursed. The spirits are on him."

I looked at the big Indian. He was over six feet tall, and

weighed more than 200 pounds, good looking and well built. He didn't seem possessed by spirits or guilty of anything. But he stood alone.

"He will die soon," Black Bear said, squinting in the sun, his face showing no emotion. "We must leave now."

The next three days were hectic. I refueled the helicopter, and helped pack equipment and food supplies on the mule trains going into the fire. I galloped the pinto from one end of the runway to the other, riding bareback because it was faster. Dub was living in the cabin, and I threw my sleeping bag in a canvas tent set up over a wood floor and porch frame. The quarters were more than adequate.

The Trapper Creek fire had blown up twice, had jumped the line twice, but the new ground crews were making progress. I wished I were out there on the line. The administration job was easy but boring—the urgency was on the smoky slopes above Trapper Creek.

One of the tasks at the guard station was to relay radio messages from the fire camp to McCall fire dispatch. And then the fateful message came.

"Cold Meadows." The fire boss's voice came over the crackling radio in the small room of the cabin. "We had a bad accident here. One dead and two injured." The caller was in a hurry.

My stomach tightened. Sixteen McCall jumpers were on that fire. I didn't want to hear a name.

"The dead man is from the Montana Indian crew," the voice crackled. "The injured are from the same crew. They've been given first aid and can walk out in the morning. The dead man was killed early today. We'll need to get the body out tonight—can you come and get him?"

"Trapper Creek," I replied, "I'll be there tonight, after dark. I'll start just as soon as we can saddle up a horse and a pack mule."

"Advise McCall dispatch," the fire boss added. "They'll want an investigation."

The wheels were put in motion. I radioed McCall fire dispatch. Hobbie Bonnet, the forester, called back and said that two Forest Service officials from Region 4 out of Ogden would fly into Cold Meadows the next morning.

At five in the afternoon, Dub and I saddled Moonshine,

and Dub threw a packsaddle over Sarge, a big, dark-brown mule. He said Sarge was the best mule for the job. I swung into the saddle, waved farewell, and with the lead rope in my left hand started up the trail along Cottonwood Creek. The trail was a gradual climb and in four hours I was at Coyote Springs, and in another hour greeted four dirty-looking firefighters just short of the fire camp on Trapper Creek. They stood in a row, seemingly guarding the dead firefighter who was wrapped in a sleeping bag, lying alongside the trail. It was almost dark. The moon would be up in an hour or two.

"What happened?" I asked the squad leader.

"A dead tree burned through at the bottom, fell across the slope and rolled downhill. Caught him full in the chest. Must have broken every rib. Two others were hit—one with a broken shoulder, and the other bruised up real bad."

"Must have ruined every organ he had," one of the firefighters said.

"Must have," I said.

The deceased was a big man, so tall his boots were sticking outside the trussed-up sleeping bag. The boots looked new, and looked familiar.

"Do you know his name?" I asked. The ground crew didn't. Another Indian man appeared out of the darkness, holding a shovel in his hand.

"He is Dan Big Bull," he said in a booming voice.

The name reverberated in the black trees. *Dan Big Bull!*

"Help me lift him on the mule," I said. Sarge's eyes were big and white in the glow of the flashlights.

"Don't cause a ruckus now," I said, talking gently to the mule. "Just easy, easy now." Dub would know how to talk to this mule. I just hoped Sarge would behave himself. The two firefighters finished tying the man to the packsaddle. I swung onto the horse and sat there a moment. The firefighters said goodbye and vanished into the darkness. They acted like they wanted to be rid of the body.

I swung the horse around in the black canyon. I hoped Moonshine could see the trail because I couldn't see a damn thing. I glanced back at the mule and the dark form tied across him. They seemed to be okay. The forest was deathly quiet. I nudged the pinto in the ribs and as we started along the trail the only sounds were the scraping of the horse's hooves.

I hoped old Sarge wouldn't spook. Dub said he was a good mule that the Fore Service had gotten from the 10th Mountain Division at Camp Hale. I had noticed the U S brand on his rump.

Moonshine followed the trail easily in the darkness. The gelding was a good mountain horse, and Dub had been working him all summer. He was taking the climb to Coyote Springs in stride. Once we reached Coyote Springs, it would be downhill all the way until the canyon opened up into Cold Meadows.

Something was bumping back there on the mule. My eyes were getting adjusted to the darkness, and as we passed alongside some pines, I saw that the man's feet were catching the branches. I retied the Indian man's feet, pulling them closer under the belly of old Sarge.

I didn't notice the blood running down from the seam of the sleeping bag, staining Sarge's underbelly and dripping onto the trail. The mule remained motionless while I retied the body.

I swung into the saddle and rode again into the night. We passed Coyote Springs, nothing more than a wide spot in the trail and a gurgling of black ice water from a spring under the pines. As we descended, the canyon walls loomed higher. The moon would be up soon.

A coyote howled behind us. I had heard the sound of coyotes many times before, but for some reason the long wavering cry made me shiver.

It was cold in the canyon and the heat from the pinto felt good. Were the wild animals of the forest also waiting for the moonrise? The eerie sounds of the night began.

A coyote howled again behind us and the call was answered by long sharp barks along the canyon walls, the sound echoing among the trees. I could see the silhouettes of the stunted firs on the ridge top, back-lit by the moon. When the first silver glow touched the jagged peaks to the left, it wouldn't be long before the trail would be washed in moonlight.

The coyote behind us howled again. Were there more than one? They sounded closer. Soon, I realized that there must have been a half dozen coyotes behind me, barking again, and this time they were answered by long howls from both sides of the canyon.

Those wild dogs of the mountains didn't go after a man. Or did they? They might go after a corpse of a man—a man on a mule, who had been cursed and couldn't defend himself. A good old Winchester lever-action would feel good about now, I thought. There was a trail axe behind me in the saddle scabbard but a lot of good that would do.

The moonrise bathed both canyon walls, and in a few minutes I would be able to see the trail down in the timber. The pinto moved along smartly, perhaps sensing he was nearing the corrals and home; or maybe it was something behind us that quickened his step.

As we rounded a small switchback on the trail, the moon appeared in its full glory. The mountains were as bright as day.

Now, I said to myself, breathing easier, there couldn't be a better job in the world than this, riding in the moonlight back in the hills of Idaho with a good horse below me and an obedient mule behind.

The coyotes behind us raised a ruckus. Would they come close enough to go after the dead man? The mule would kick if they did. Or he could bolt, charging off into the timber, the corpse bouncing up and down on his back until something awful happened.

There were at least a dozen coyotes behind us, and a half dozen on each side, judging from the howls, yips and barks. Was my imagination running away and exaggerating the numbers? Sure wish I had that Winchester with the tubular magazine full of cartridges.

The canyon opened up, the steep walls giving way to rolling timber. We were getting close to Cold Meadows—just a couple more miles and Dan Big Bull would be safe in camp. Maybe he *was* cursed. I wondered what the curse was. Was it continuing now?

We had entered the big open meadow, the grass wet with the early morning dew and sparkling in the moonlight. The cabin and tent were just up ahead. I'd wake Dub and get him to help unload the body. Our small pack string stopped alongside the wooden porch of the tent. The meadow behind the tent was ringed with a line of black timber, and an owl hooted in the trees. The coyotes, sensing the station and smelling the horses, backed off into the edge of the forest, biding their time.

I swung down from the saddle. My watch showed three o'clock in the morning. Dub came up and we untied the man and laid him onto the wood porch of the tent. The last of Dan Big Bull's blood poured out of the sleeping bag and over the porch.

"His chest and lungs were crushed," I told Dub.

"The motion of being packed on the mule must have pumped his blood out," Dub said.

"That's why the coyotes had us surrounded."

"You get some sleep," Dub said. "I'll take the stock to the corral." Moonshine and Sarge needed no urging as he led them to the corral for oats and water.

I fell onto the sleeping bag and was out in an instant. But not for long. Outside on the porch there was a terrible ruckus. Snarls, barks, and howls filled the night air. The coyotes had come back! The screeching propelled me off the bunk and I ripped open the tent flap. There on the porch were a half-dozen coyotes, jumping, snarling, and tearing pieces of the sleeping bag to shreds. White fluffs of down filled the air.

"Get out of here! Go away!"

"What in the hell is going on up there?" Dub carried a lantern up the walk from the cabin.

"The coyotes are trying to eat the dead man!" The tails of the wild dogs could be seen bouncing away in the moonlight.

Dub came up with his dog, a brown curly-haired little fellow that wasn't more than a foot long and weighed about ten pounds.

"We'll keep those sons-of-bitches away," Dub declared, and tied the little dog to the corpse.

"Think he'll scare them off?"

"He'll keep them away," Dub said, and picked up the lantern for the walk back to the cabin.

The little dog was brave. He looked longingly after his master, but he was securely tied to the bloody corpse, and he shook so hard the floorboards of the porch rattled. He'd have made one good mouthful for a hungry coyote.

The coyotes never returned. In a few hours the day dawned bright and clear, and Dub woke me with a telephone call from Hobbie Bonnet.

"Two top officials from Ogden will fly into Cold Meadows today about noon to investigate," Hobbie said. "I heard you brought the dead man out last night."

"Yes. His name is Dan Big Bull."

"Well, clean him up so he'll look good for the investigators."

"I'm no undertaker, Hobbie," I said, picturing the bloody, torn sleeping bag. "Those two guys don't want to see that fellow up close any more than I do."

The Noordyne landed at noon. The two officials, in clean and pressed Forest Service gabardine, were on their way to the fire camp, and they didn't act like they wanted to look at anything or talk to anybody anyway. Dub and I loaded Big Bull's body into the cabin of the Noordyne. The two officials would have to step over the bloody man in the sleeping bag when they got back into the plane, I thought.

I latched the cabin door and walked back to the guard station. On the way, I stopped at the corral to check on old Sarge. The big mule looked none the worse for wear, but the whites of his eyes were large, and he rolled them first to the right and then to the left. Then they seemed to glaze over. I rubbed his nose for a few moments and talked gently to him, but he didn't notice. He was staring at the line of dark pines behind the tent, seeing something there that only an animal could see.

The body of Dan Big Bull was taken to a local undertaker and then transported back to his tribe in Montana. The two injured men were walked out to Cold Meadows and flown by helicopter to the hospital in Cascade. They recovered from their injuries and were back on the fireline the following summer.

But it was a different story for Sarge, the mule. Val Simpson, the ranger, said he was never the same after that night.

Chow hall at McCall

Chapter Thirteen

TO SALMON AND BACK

"I can see that you're a jumper
and not a common bum,
by the way you hold your coffee
and stir it with your thumb."
Anonymous

In 1954, the small eastern-Idaho town of Salmon could boast of having thirteen bars and two houses of ill repute. All thirteen bars were on Main Street, just off the river bridge. The two houses of pleasure were off Main by one block, tucked along the short side streets near the river, placed there, perhaps, to be out of sight of the city mothers and fathers, who were well aware of their activities.

Salmon is perched in a high mountain valley, straddling the Continental Divide just west of the Montana border. In 1805, the Meriwether Lewis and William Clark expedition, searching for a route west across the Bitterroot Mountains, lingered awhile along the river. In 1819, fur traders made the first permanent white man's camp along the banks of the Salmon River, preferring the warmer valley to the snow-filled mountains. By 1840 the fur trade was over, and in 1850 several Mormon missions were built in the vicinity of what is now the town. But the occupants were run off by the Shoshone Indians, who, the archaeologists estimate, had already lived in the region for 8,000 years. For decades in the nineteenth century, this beautiful valley lay uninhabitated by white settlers. Even the old immigrant road missed the valley, coming down the Madison River east out of Montana and linking up with Henry's Fork in southeast Idaho. The road then joined the Lander Road and took a westerly course to Oregon.

Gold was discovered in 1866 along the Salmon River, and a rapid parade of inhabitants soon followed, solidly establishing the town in 1867. Through the late 1800s the town of Salmon grew, and ranching followed at the turn of the century. In the 1930s and 1940s, the town became the gathering place of cattlemen, miners, loggers and sheepherders. Many of these boys, after spending months in the high country, were well entertained by the ladies in the small houses just off Main Street.

The direct distance from Salmon to McCall across the rugged central primitive area of central Idaho is 115 air miles, but to drive it means a long go-around and takes a full day, and depending on the time of year, can be as much as 400 miles. You could ride a saddle horse to McCall, following the old trails across the rivers and over the mountains, but you'd need a string of packhorses and most of the summer.

The Salmon River starts at an elevation of 10,000 feet in the rugged Sawtooth Mountains high above the village of Stanley. This beautiful river, full of salmon and steelhead trout in the first half of the century, flows north from Stanley. The river picks up the cold water from the Parsimeroi, and then the Lemhi River, as it flows under the river bridge in downtown Salmon, just a stone's throw from the painted ladies in the little houses. From the town, the river wants to get away from civilization again and it continues north, then turns west to bisect the state of Idaho. A narrow dirt road ends at Corn Creek and from there only high canyon walls accompany the river as it flows through the wilderness for eighty miles. Eventually, it sweeps out through the breaks in the Seven Devils country and into the Snake River. The Salmon River was named by the settlers the "River of No Return." It drains some of the wildest, most beautiful and untrammeled timbered country on the North American continent.

The headquarters of the Salmon National Forest was—and still is—located in the town of Salmon, in Forest Service Region Four. The smokejumpers in McCall had the responsibility to put out fires across this enormous mountain country.

Late in September of 1954, the fall rains had not started and the woods remained tinder dry. Fifteen jumpers remained on duty at McCall, the others having returned to school or to their winter jobs. Shortly after the noon hour on September 20, the buzzer rasped eight times. The next jumpers up were Spence Miller, Len Walker, Gene Marker, Carl Shaver, Pat Daly, Frank Kerby, Ray Beasley and myself. Paperlegs Peterson was the spotter. Fifteen minutes later, suited up, we climbed aboard the Ford. The afternoon was clear and bright, but the airplane ride was rough and rocky from the winds coming up the canyons. We pitched back and forth in the cabin, not having seat belts or even seats to hold us down. The air turbulence was severe—the firepacks in the back of the cabin shifting, tool handles wobbling back and forth—when the plane hit a downdraft. None of us said much above the roar of the three engines. Lunch at the mess hall had been a big one of hamburgers, mashed potatoes and pie; we were hoping it would stay down. Two hours later the Ford Tri-Motor was circling a fast-moving fire on the headwaters of Mustang Creek on the Salmon National Forest.

There were long slopes here, green and wide and gentle, grass and sagebrush, and stands of tall timber on the tops of the mountains. From a distance the timber looked like dark green carpets artistically placed on the tops of the mountains by a skilled hand. The long slopes below the timber were large enough to jump into, and didn't look excessively steep. The silver sage on the open slopes below the timber sparkled in the afternoon sun.

The fire was still on the ground, where we wanted it to be, the flames licking away at the brush below the big ponderosas. Tall Douglas firs grew next to the jump spot; swirling in the wind, they appeared to be watching the smoke. The pines and the firs were more than 100 feet tall, old virgin growth, I thought, been there for centuries. We'd put the fire out before the woods were torched into a stand of blackened snags.

"Piece of cake!" Spencer Miller shouted, looking out the jump door. And it was, except for one thing: the unstoppable wind. Spencer would be the first jumper out. He turned and faced us. We had all shifted to watch the smoke and try to confirm in our own minds Spencer's optimistic opinion.

The wind howled from the canyon below and over the rounded hills, laying the smoke flat across the mountain. The big plane shuddered as a mass of air caught the broad metal fuselage. Wings that looked stout enough to withstand a hurricane vibrated, rocking right and left. The three engines screamed as the plane dropped with the wind, an invisible wave of air flowing downhill from the ridge. The pilot had the throttles wide open and the propellers flashed in the sun, searching, grasping for a hold in the sky.

"Too much wind!" Paperlegs shouted. He was aft of the open door facing us, his feet spread against the buffeting of the plane. Maybe he was right. This would be the final jump of the summer for me and I sure as hell didn't want to break a leg on the last fire.

Paperlegs Peterson looked again at the smoke streaking over the top of the mountain. "Besides," he shouted, "we're over 9,000 feet and the air's thin. It's an oven in the canyon. You'll drop like rocks."

Paper would make the decision, but then Gene Marker yelled, "Throw out the drift chute, and time it!" Jim Larkin turned around in the cockpit. Paperlegs circled his hand for a drift-chute pass and indicated by touching his watch he would time the descent. Jim banked the wings for a straight pass above the hillside, steadying the aircraft in the turbulence.

Ray Beasley was my jump partner on this two-man stick and we had planned to take pictures of each other on the way down. We pulled the cameras from our jackets and began adjusting the shutter speeds and lens openings.

"Smile for the camera," I shouted at Beasley. "But don't get too close."

"This is going to be fun," Ray said, rolling his eyes. With this group we already were assuming we were going to jump.

Nine of us watched the drift of the chute and made mental checks of its descent time. It landed 300 yards downwind of the jump area and took less than a minute to fall. "There's too much turbulence and no air," Paperlegs repeated from the door, his blue eyes watering from the prop blasts.

"We want to jump, Paper!" Spence Miller, Gene Marker and Carl Shaver shouted. Spence had pulled on his helmet and snapped the chin strap, as if the decision had already been made. Ray and I continued to fuss with our cameras. Frank Kerby inverted his football helmet and put it up to his mouth, as if he were about to lose his lunch.

"We didn't come this far to turn back," Carl Shaver said, coming alive to the situation. We all shifted our positions again to look at the landing spot.

"Yeah, Paper, we're all second-year and third-year guys," Marker said, grinning as usual.

Paperlegs looked us over. We *were* all old hands, except Frank Kerby. This was his second year, but he was sick as usual. He sat, his helmet in his lap, gray eyes bloodshot, and face the color of typing paper. The guys all wondered why Frank even wanted to be a jumper; he'd get airsick every time up. It must have been torture for him. He was a quiet guy who worked hard on the line, but he wasn't close to any of the jumpers. Maybe he was trying to prove something to himself. It seemed obvious that one way or another, Frank wanted to be out of this plane. Paperlegs didn't consult him as he looked at each of us.

"It's *your* legs, you idiots," Paper muttered, and unhooked the safety strap from the open door.

Spence Miller and Len Walker were up. They disappeared and left only the shrieking propwash in the empty door. Paperlegs pulled the two flapping static lines into the cabin and we shifted again to watch the parachutes. The plane banked and came around over the jump area, 1,000 feet up. Both jumpers hit hard, and neither moved. At that altitude they looked like crumpled cargo bags.

"Marker and Shaver, you're up," Paper bellowed. "See what happened to those guys. And damn it, do good rolls!" On the jump pass, Paperlegs slapped Shaver's thigh and he and Marker went out, like they were glued together. We watched the two parachutes parallel each other until they got smaller and smaller. They hit the ground and like the first two, Marker and Shaver did not move.

Ray and I made up the last stick, and we waited behind the bulkhead. Pat Daly and Frank Kerby were next. Frank was pale, and I think he would do about anything to get back on the ground. "Somebody is up!" Daly shouted. Pat Daly, five-foot eight inches tall, almost bald at the age of twenty, was a cool stud. He'd watch

Frank on the way down. On the ground, two men stood, moving as if in a coma, spreading out their orange streamers on the ground.

Beasley tapped me on the arm. "Let's go into the trees. Two of them haven't moved." Ray usually didn't get worked up. "We'll take to the trees at the top of the clearing."

"Into the trees," I shouted back. "And take some pictures."

Daly and Kerby landed on the open hillside fifty yards uphill from the first four. They rolled with the wind and stood up, waving their arms. Four jumpers were now running to assist the first two.

"You're up," Paperlegs said. "Watch yourselves." I put my foot on the outside step, hands on the sides of the door. Beasley jerked on my riser snaps and static line for the equipment check, second time. He was ready to follow me out. He also had one gloved hand on the edge of the door, squeezing in tight behind me. "Into the trees," he shouted again into the ear hole of my helmet. I nodded, but my face was in the slipstream and I was concentrating on the green clearing coming toward us 1,000 feet below. The pilot brought back the power on the three engines and the roar settled to a rumble. "See you on the ground, Kid," Beasley shouted.

Paperlegs slapped my thigh hard and I stepped out. The snap of the canopy opening hit me, and the chute blossomed against the sky as it was designed to do, a miracle that always amazed me. I stopped the swinging of the chute by pulling on the risers, then reached inside the collar of my jacket for the Argus. I tugged on the right guideline to turn and look for Beasley. There he was, outlined against the blue sky, slowly swinging and taking pictures. I pointed the camera at him and snapped off two frames.

Something was wrong. The wind through the face mask didn't feel right. The smoke from the fire had rolled over and changed direction ninety degrees. The wind blew up the ridge instead of across it. This could get hairy. Bease and I had decided to land in the trees, but not over the mountain and into the next drainage. I tucked the camera away, turned with the guideline, pulled down on the risers and planed into the wind. Ray had figured it out—I could see him off to the right, doing the same maneuver.

The air *was* thin and I fell fast, the ground and trees coming up quick. Tall firs below slid past as my ground speed accelerated with the wind. There were jump stories about guys landing in trees, the branches spilling the air from the nylon, and the parachutist free-falling alongside the tree, breaking arms and femurs.

The instructions were to avoid grabbing the top of the tree as you came in, but I grabbed the small top anyway as I crashed into the branches. It was a survival reflex, trying to hold onto something, anything, and it could work if the canopy didn't pull me off. The top of the tree was only three inches thick where I grabbed it but I was stopped. The red-and-white chute buffeted, stalled, lost air and swirled down alongside the branches below. It could have been real bad if the canopy had kept going and I was left clutching the top of a 150-foot fir. The safe landing had been just plain luck.

A pack of Luckies in my shirt pocket flew in all directions when I hit the branches. In the middle of this display, I saw Beasley land in a tree seventy-five yards uphill.

Now I had two problems. The first was the thick resinous smoke drifting through the branches. The wind changed again and it gusted toward me. If the fire blew up, this was the wrong place to be. Second, I was in the top of a tree with branches so thick I couldn't see the ground. I'd have to depend on the hundred-foot rope, and the tree was higher than that. I went through the let-down procedure, tying one end of the rope to the risers, taking pressure off the D-rings, unsnapping them, then unhooking the chest and crotch snaps and stepping out of the harness. The rope was secure and I dropped the coil and watched it disappear into the branches.

The smoke got thicker, and I thought I could smell the heat in it. With one hand on the rope and the other holding onto the branches I climbed down, like descending a ladder. The trunk became thicker, first two feet, then three, and finally I could see the ground. Beasley walked across the forest floor, stopped at the base of the tree and smiled; the end of the rope dangled ten feet from the ground. I dropped easily to a soft landing in the pine needles.

He pulled out his pipe, and without lighting it, stuck it between his teeth. "That was different, wasn't it," he said.

"We were dropping like freaking rocks."

"And moving sideways at the same speed," Ray said. "Get any pictures?"

"I took two of you, then saw the change in the wind and had

to go to work."

"What are these cigarettes doing all over the ground?" Ray said and laughed.

We joined the others at the fire. Miller and Walker had hit so hard they were stunned. It was the same for Marker and Shaver. There were no broken ankles or legs, and they were fine after collecting their wits, so we gathered the shovels and pulaskis and went to work digging line. The firs and pines were tall and thick-barked and the fire stayed on the ground. We dug a hasty line around the perimeter and watched the winds.

A long three days later, after shoveling a wide perimeter and checking for remaining smokes, we declared the fire officially out. Ray and I retrieved our chutes, still hanging in the branches, and they came loose with hard tugs on the let-down ropes.

"You were lucky you didn't fall out of that tree," Ray said.

"I grabbed the top, after thinking about the jump on Mosquito Peak when Chip Beatty fell out of that yellow pine."

"Broke his leg in three places."

"Chip said his jump career was over."

"He told me he was going back to work in a lookout tower."

"Now that he's married," Ray added, stuffing his parachute into the elephant bag. "His wife wrote the book, *Lookout Wife.* Before Chip became a jumper, they spent the summer in a lookout. Maybe I'll read it someday," Ray said, his interest fading with both the thought of marriage and living in a lookout. Ray was more the social type.

The hike off the fire took us down an open ridge to the canyon of the main river, and a Forest Service truck waited for us at the mouth of Corn Creek. For a few minutes we watched a packer rope up a string of mules to ride back into the burn and bring out our equipment. We told him about the fire and where to find the stacked firepacks and said he could have the cans of food we left. The mules snorted like they wanted to be off. We quickly climbed in the back of the truck. We had a long ride to Salmon but it was Saturday night and the town was waiting.

It was dark when we arrived, hot and dusty, in Salmon. We told the Forest Service driver to drop us downtown and we would see him later. The Crescent Club saloon on Main Street was the hottest spot in town. We walked through the swinging door and threaded our way in the smoky haze to two tables on the dark side. A western band was positioned on a stage in one corner: a pretty blonde vocalist with painted-on jeans and a pink sequined shirt, a guitarist wearing a Stetson, a guy on the bass and another on the drums. Once our eyes became accustomed to the place, we could see that the cowboys had all the girls cornered.

Several of our guys had bought some tequila and bourbon and kept the spirits in paper bags under the table. Len Walker drank the tequila straight and chased it down with bourbon. Spencer Miller, without question as tough as Walker, drank the bourbon straight and finished it off with beer. Spencer's goal was to make the airborne infantry a career (which he did, serving two tours in Viet Nam and becoming a Lieutenant Colonel), and I thought for awhile that night he was going to start his training right then and there. The rest of us drank beer from long-necked Lucky Lager bottles.

The saloon was filled with the Saturday-night crowd from the ranches and hills: cowboys, sheepherders and miners. The cowboys had the entire female population corralled. Having spent the summer in the mountains, the herders weren't sure just how to act in town. The hardrock miners, older and wiser, were in the bar to drink whiskey and watch the action.

Even if the cowboys had all the girls, it didn't slow down Len Walker. Broad shouldered, dark haired and good looking, twenty-year-old Len stood over six feet in his jump boots, and had hands on him that made an axe look like a play toy. Len grew up in logging camps in north Idaho, and at the age of eighteen decided to go to forestry school, unless something came along that was more adventurous.

Len's philosophy was that there was only so much time in life and none of it was to be wasted. He put down his drink, walked over and introduced himself to the cute brunette sitting at a table with five cowboys. She had short hair, red lipstick and hand-fitted Levis, a lovely senorita. The cowboys in astonishment pushed back their big hats when she eagerly accepted Len's offer to dance.

On the floor the two of them flowed to the music. Len had the charms; the young woman placed her arm behind Len's back, moving it upward. Len pressed her against him, speaking soft words in her ear. When the rhythm of the dance was just right, she placed her leg behind Len's. The cowboys glared.

When the music stopped, Len escorted his pretty new partner back to her table, and in a manner so smooth and in words so soft, said thanks, and you would have thought maybe he was from Harvard or Yale, instead of being raised in logging camps and hooking chokers to logs at the age of fifteen.

"She's cute—I'm going to ask her again," Len said, turning his chair around and sitting on it backwards just so he could watch her.

"You're going to get in trouble," Beasley advised.

"Her name is Rita, and she likes me," Walker said, ignoring us.

Gene Marker and Carl Shaver had cut out several pretty ladies and were smoothly moving with them to the music.

Len went to the table with the cowboys and extended his hand again to Rita. She smiled, slid her chair back and put her arm around Len's waist. It was easy to see she liked him. All the girls did.

Len and Rita took the dance one step further. The music was a lively tango. Rita's earrings flashed in the light as she tapped her way to the beat. Len grasped her waist with those big hands of his and lifted her atop the bar. The patrons picked up their beers and leaned back, and Rita danced from one end of the bar to the other. Len lifted her down to the floor, and without missing a beat, they again moved to the music. Everyone in the saloon was watching them...they were good. The cowboys were glaring harder than ever; these things didn't happen in *their* bar, with their women.

When the music stopped, Walker led Rita once again to her table, motioned to the bartender to send her a drink, and angled to the bar for another whiskey.

The cowboys at her table tried to figure out what to do next. While Walker was at the bar, one of the tall guys, maybe Rita's boyfiend, came over to our table. He was a working cowboy, dangerous-looking but quiet, his hat sweat-stained, the edges rolled up in front and back. I knew what was coming.

"Maybe you better ask your friend not to dance with my girl," the cowboy said.

"We can't control Walker," Spence Miller said. "Sorry."

"He's an animal," Gene Marker added.

Walker returned to Rita's table, arriving at the same instant as the cowboy, who slid between Walker and Rita. She was already standing and it was plain to see she wanted to dance with Len.

"What are you drinking, partner?" Len asked the tall cowboy. Len wore a steel-trap smile, just like Bertram's. I guessed that Walker was as fast as Bertram, maybe even faster. Without waiting for an answer, Walker reached down and picked up the cowboy's drink and downed it in one quick swallow, not taking his eyes off the cowpoke for an instant. Rita slipped away from the table.

The cowboys and the sheepherders were the real thing—work-toughened men who could obviously scrap. It was three to one against us if anything got started, but we were all as tough as cinders from running the ridges and working the line all summer.

Walker was pushing it now. He slowly lowered the glass to the table, and without taking his eyes off his competition, grabbed the next cowboy's long-necked beer and swept it in an arc to his mouth. Or was he going to break it over the cowboy's head?

That was too much for the cowboys. They all jumped to their feet, as if on command. Was Walker nuts? Did he think he could take them all at the same time? The drinkers at the bar nudged each other to turn around. Things were about to get western in Salmon.

The front door of the Crescent Club burst open and the sheriff and three deputies charged into the saloon. They pointed over at us, sitting around the tables. Why us?

"Boys," the sheriff said, right hand on his big horse pistol, "you have two choices. The first is to spend the night in jail, and the second is to spend it in the Forest Service warehouse. In either case, we have two cars waiting outside to take you wherever you decide."

Why screw up a good summer? "We'll take the warehouse," Spence said.

"Didn't mean to cause any trouble," Marker said. "We were

just dancing."

"That's good," the sheriff said.

The deputies drove us to the warehouse, and without so much as a farewell, dropped us off. They were very serious young men.

Somebody brought out more tequila, bourbon and beer. The party continued, amid rows of firepacks and neatly stacked sleeping bags. Len Walker was just getting warmed up, and he said he knew a few of the ladies down by the river. We borrowed a Forest Service supply truck, and soon we were headed to Mr. Smith's House of Pleasure, as Len called it. The price was five dollars and Len had to borrow the money from us, because his cash, he said, went to supply the tequila.

Old Mr. Smith, wearing a starched white shirt and no tie, met us at the door. This was his big night: eight guys, just off a fire. He welcomed us into the parlor and poured us all a tall glass of whiskey. Now then, Mr. Smith said, who will be first? The only one of us so inclined was Len, and he enthusiastically followed Mr. Smith down the hall, the change jangling in his pocket. He was back in what seemed like minutes, and none the worse for wear.

"Well, who's next, gentlemen?" the old fellow asked. He poured another round of drinks, waiting for a reply. Our eyes were down, looking at our drinks. The old man's face fell a little.

"Who's next?" A husky female voice came from the hall.

"I'm next!" All eyes looked up. It was Len again. "I need another five dollars."

We dug deep, some offering a quarter, some a silver dollar, until once again Len had five bucks. He gave it to the old man, who smiled just a little.

"He's in love again," Marker said.

We spent the remaining hours of that Saturday night in the warehouse, finishing off the liquor. The Ford Tri-Motor landed at nine o'clock the next morning at the Salmon airstrip, a quarter of a mile from the Forest Service station. The skies were bright and clear when we climbed into the plane for the trip back to McCall.

On the return flight, Frank Kerby, as usual, got sick. He had been in good spirits the night before and had drunk as much as the rest of us but once in the air his face turned from parchment white to green, so vivid that Ray took color pictures of him with his Leica. Twenty minutes into the flight, Kerby got sick in his helmet, threw the contents out the window into the slipstream, and they blew back and sprayed Walker, Miller and Marker, sleeping on the firepacks in the back of the cabin. Kerby was ordered to sit for the rest of the trip, and as standard practice on all trips afterwards, in the back of the plane.

Oh well—it *was* the end of summer.

McCall jump base, 1959

Chapter Fourteen

A THOUSAND SUMMERS

October had almost arrived and it was time to go home. I planned a two-day stop in Boise before taking the train east, but the Cougar Creek fire blew up twice and we spent two extra days knocking it down. I had already missed registration at college and I'd have to start classes a week late. The district ranger wrote a letter to the school authorities explaining that forest fires had detained me and my services were needed on the line. The professors at forestry school would understand. A copy of the letter was in my pocket for safekeeping. The two-day stop in Boise was to see Liz. She said she'd be waiting, but that was a week before, and the fire kept exploding at Cougar Creek. I hoped, even prayed a little, that she'd still be there.

I thought about the trip to Big Creek three years before when I met Liz. We tried to spend as much time together as we could, but events of the summers kept interrupting. She was either at Boise or Big Creek, and I was in McCall. We liked each other more than a lot, but we were young, and life wasn't meant to be serious. She didn't have a car, nor did I, but somehow I'd borrow one and drive to Boise when I could, which wasn't often when the fire season started. Our time together was always great: dancing at the Miramar, having pictures taken together at the State Fair, the hot springs and more dancing at Idaho City on the long weekends. Back at school, we wrote to each other weekly. Maybe Liz was my girl of the summer, and I was her guy.

It was a warm afternoon when the green Forest Service pickup bounced across the dirt parking lot behind the supervisor's office in McCall. Four jumpers leaped from the back before the truck rolled to a stop. Reid dropped off the fire report at the fire dispatch office. I double-timed to the phone inside the building. I *had* to see Liz before going east. As I crossed the yard of the Forest Service station I thought back to the day in late May three summers ago when I walked through there with a duffel in hand, and Marker and Shaver drove up in an open Jeep and said hello. A lot had happened since that day … a lot of great memories.

Two weeks earlier, I had asked Liz if I could see her again before school started. She said, "Yes, call me when you get back into town." Maybe she thought I had already left for the east; back in the mountains there wasn't any way to let her know I'd been delayed. I said a quick prayer that she would be home as the phone rang. Liz answered, excitement in her voice. "I'll be in McCall Friday evening. Bob Fogg is flying me into Big Creek Saturday morning. He's taking supplies into the outfitters. I'm going to help Mom and Dad for the hunting season. Can you meet me Friday night at the Shore Lodge?"

I felt myself blushing, even over the phone. This was better than meeting her in Boise. She was coming *here*. Tomorrow night. I thought I was in love with her.

"Of course. What time is good for you?" I answered, trying to remain cool.

"Seven-thirty in the lounge. I'd like to see you before you hit the books and start chasing those good-looking coeds," Liz teased. Her voice was smooth and laughing and carried a song of adventure and mystery. Was it because she was meeting me? Was it the flight into the backcountry and spending a gorgeous fall there? Was it youth and adventure? That's why I liked her so much—she was beautiful *and* adventurous. Or was it something deeper, something more serious, a feeling neither of us could explain?

"I'll be studying hard."

"I'll bet you will. Seven-thirty then, tomorrow night." Liz hung up.

The Shore Lodge, an elegant wooden structure with nice

rooms, a restaurant and a bar, was on the south edge of the lake, an easy walk from the base. At that time of year the small town was quiet: the summer people had returned to the valleys, and the bars were inhabitated only by a handful of loggers, a half-dozen smokejumpers and several foresters. In three months, the foresters would be marking timber on snowshoes. Now those last days of September were filled with crystal-blue skies, and above the mountains, piles of white clouds. The aspens on the high ridges in the backcountry were turning yellow.

I put on a clean shirt, faded Levis and a pair of comfortable western boots. I could feel my shoulder muscles under my shirt, muscles I had gained on the fireline. White shirts and ties were required in the eastern social world and I'd be wearing them soon enough.

At that moment I felt so good I thought I could take on the world. It was the last day at camp for me, and for a moment my thoughts went back to the jumpers. I would miss them all. They were all tough guys. There were a lot of skirmishes ahead for me in the world, but I knew I could handle them. Carl and Bus and Paperlegs: there would be times years later when I could have used their help. They weren't there, of course, but they had trained me well, and I think their spirits were there. Sometimes I thought I could even hear their voices, telling me what to do.

It was seven-thirty when I walked through the lobby and into the wood-paneled bar. The lounge was almost empty, filled only with the quiet and calm feeling of late summer. A couple sat at a secluded corner table. A female vocalist adjusted the microphone, her accompanist at the piano. Huge windows on the lakeside framed the big ponderosas and the purple mountains in the distance. The sun was low in the sky, starting to drop behind the far peaks. As I walked in I could see the waters of the lake, a deeper blue than usual, reflecting the evening sky of late September.

A beautiful young woman sat on a tall chair at the corner of the bar. She wore a white sleeveless blouse, yellow shorts and white shoes. Liz was five-foot seven, and had the most slender and well-tanned legs I had ever seen. As she turned, the late evening sun through the window touched her hair.

She smiled. "I missed you," she said.

"And I missed you."

The vocalist began her song: *"If it takes forever, I will wait for you. For a thousand summers, I will wait for you."*

"Remember," I said, not taking my eyes off her.

"We danced to that song late into the night at Idaho City," she said, and lightly touched my leg.

Those brown eyes taunted me with a mischievous look. And that moment, as she smiled and turned toward me, I remembered for the rest of my life.

The sun slipped below the mountains, and outside the remaining light played out among the branches. A breeze came off the lake and drifted through the trees. On the other end of the universe a few faint stars appeared. I leaned forward, put my arm around her waist, and kissed her.

The world was young. We were young. The evening was just beginning.

The vocalist sang, *"For a thousand summers, I will wait for you."*

"For a thousand summers," Liz said.

Wildfire aftermath

Epilogue

The next day I caught the train east, registering just in time for classes. I never saw Liz again. Time and too many miles separated us. She moved from Boise, became an airline stewardess and married a pilot.

After graduation I served in the 10th Mountain and 9th Infantry Divisions, and during those years I met Ellie. She was—and still is—a petite raven-haired beauty with magnificent hazel eyes. She stunned me then, and still does. It all worked out for the best.

I didn't fight fires for a thousand summers, but I envied the guys who did. In the towns and watering holes of Idaho, I still run across those jumpers today, the boys who leaped out of airplanes in the fifties. They are the best people I've ever known.

Each fall when I returned to forestry school, my friends would ask why I wanted to be a smokejumper, what it was like. Most of them, as students in the profession, had worked for the Forest Service in the West, and they had fought fires. They walked in on the ground and I had come in from the sky. I found it hard to talk about my real feelings, but after several rounds of beers, the stories started. We talked about the great conflagrations, the running crown fires we had battled in Idaho, Montana and Washington. We exaggerated on the sizes of the blow-ups, the fires that had incinerated whole forests in a matter of hours. We expounded on the sweat, dust and heat of digging line day after day; and of queuing up at the smoky fire camps, waiting for the servings of bread and beans. We lied about the women and the barroom brawls.

But the hardest part to express was the love for the sky and the mountains—the pure joy of soaring over granite peaks, of smelling the exhaust smoke of airplane engines and flying across sun-drenched ridges, of turning in the door, giving a thumbs up to your jump partner, and then stepping into space above a timbered canyon.

I also wanted to tell them I believed in the immortality of man, and in God, and that I knew that up there somewhere in those high and windy places the eternal spirits dwelled. We had all been there. In some ways, we are all still there.

PARALLEL
Pueblo Summit
Werdenhoff Mine
A
BM 6236
Wolf Fangs
Wolf Fang Peak
9007
BM 6679
Mine
Middle Fk
Smith
Mc Crae Mine
BM 7519
340
Round Meadow
371
Independence Mine
Elk Summit
BM 8670
Fawn Mdw
Golden Cup Mine
Goldman Cut
Crystal Spring
Windy
Ridge
Mount Eldridge
ELK
9207
Government
Landing Strip
BIG CREEK WORK CENTER
Hogback
Big Creek
Dixie Mountain
9070
Ludwig Mine
BM 6418
BIG CR AIRFIE
Edwards
USMM 2
Sunday Mine
B
343
Greeley Mountain
USMM 2866
Moscow Mine
BM 5704
Logan Mountain
Moore Point
8971
Logan Lake
BM 6952
RD
BM 5888
BM 6918
BM 7605
Profile Gap
Willson Mine
Glascow Mine
Crater Peak
USLM 1
Red Metals Mine
Crater Lake
Middle Lake
Lotspiech Mine
BM 7028
Coin Mountain
9065
Goat Mountain
8955
PINNACLE 9273
Ridge